# The Natural Laws of the Universe

## Understanding Fundamental Constants

Jean-Philippe Uzan and Bénédicte Leclercq

# The Natural Laws of the Universe

## Understanding Fundamental Constants

Dr Jean-Philippe Uzan
CNRS – Institut d'Astrophysique de Paris
Paris
France

Dr Bénédicte Leclercq
Pour la science
Paris
France

Original French edition: *De l'importance d'être une constante*
Published by © Dunod, Paris 2005
Ouvrage publié avec le concours du Ministère français chargé de la culture – Centre national du livre
This work has been published with the help of the French Ministère de la Culture – Centre National du Livre

Translator: Bob Mizon, 38 The Vineries, Colehill, Wimborne, Dorset, UK

SPRINGER–PRAXIS BOOKS IN POPULAR ASTRONOMY
SUBJECT *ADVISORY EDITOR*: John Mason B.Sc., M.Sc., Ph.D.

ISBN 978-0-387-73454-5 Springer Berlin Heidelberg New York

Springer is a part of Springer Science + Business Media (*springeronline.com*)

Library of Congress Control Number: 2008921345



Cover design: Jim Wilkie
Copy editing and graphics processing: Alex Whyte
Typesetting: BookEns Ltd, Royston, Herts., UK

Printed in Germany on acid-free paper

# Table of Contents

# Foreword

Constants are the reference values of the physical sciences. They embrace and involve all units of measurement, and orders of magnitude, of the various phenomena and the way they relate to one another. They may be seen as the pillars of physics. This is especially true of the three principal constants – the speed of light $c$, the gravitational constant $G$, and the Planck constant $h$ – since they underpin three fundamental theories: special relativity, Einstein's gravitational theory, and quantum physics. These three constants represent the three dimensions of the so-called 'cube of physical theories', a construction that sums up the state of modern physics. They also furnished George Gamow, the famous physicist and renowned populariser of science, with the initials of his creation, C.G.H. Tompkins, who explores universes where constants are fundamentally different from those we know: when the speed of light is very small, he sees those around him flattened in the direction of their motion. The variation of these constants would modify our universe and the laws of physics, which govern the objects around us. However, certain theories, such as string theory, predict the variation of these constants in time and space. Can we measure these variations in order to test these surprising theories? Until now, no variation has been, unambiguously detected to the satisfaction of all, although this is not for the want of trying! Atomic clocks have established certain constants to 15 significant figures(!), while meteorites allow us to go back in time to the birth of the solar system. Going much further back in time and deeper into space, astrophysicists are examining light from very distant quasars. This has revealed, in certain cases, a possible very small variation that has not, however, been confirmed by all observers. Our own observations in the millimetric domain, some in collaboration with Swedish scientist Tommy Wiklind, do not detect this.

Jean-Philippe Uzan and Bénédicte Leclercq have written their book in the form of an 'official inquiry' into the importance of a possible variation in fundamental constants. Guiding us though the history of the ideas of physics, they evoke major discoveries, from Galileo and Newton to Planck and Einstein, and, finally, discuss current questions raised by ever more accurate observations. Approaching physics by way of its constants is to distinguish the fundamental from the particular, and to recognise different physical forces, which we are unable to draw together into one unique force, the ideal for those seeking a unified theory. The development of theories leads to simplification, analogy and the unification of phenomena. Physicists seek to explain why the world is as it is, but why can they find no explanation for the value of the mass of an elementary particle such as the electron and the proton? Can we have confidence in the promising theory of superstrings, which would reinterpret those particles as states of vibration of strings, extended objects appearing as particles only in those macroscopic dimensions that we can appreciate? That would indeed be a novel view of our universe...

This highly instructive book is complete and coherent, and goes beyond the subject of constants to explain and discuss many notions in physics. Along the way, we encounter some exciting details: for example, how scientists discovered the existence of a natural nuclear reactor at Oklo in Gabon, although for months, the secret service had suspected other causes...

The reader will find this an agreeable and mind-expanding book, with many an anecdote from the history of scientific discoveries. This is science as it is happening, not what we hear in lectures!

*Françoise Combes,*
*astronomer at the Paris Observatory and Member of the Academy of Sciences.*

# Acknowledgements

The authors are grateful to Françoise Combes, who honoured them with her preface to the text, and to Anne Bourguignon, who suggested that they write this book: the pleasure they had in writing it is, they hope, shared by its readers.

They offer warm thanks to Roland Lehoucq and Christophe Poinssot for their expert and attentive proofreading, and to René Cuillierier for having produced the first popular article in French on the subject. Finally, they thank the many colleagues who have debated these questions, and have thrown so much light on them through their reflections and their knowledge in recent years.

# Introduction

## CAT AMONG THE PIGEONS

Early in 1999, a brief communication appeared in the prestigious American scientific journal *Physical Review Letters*, and created a stir throughout the world of physics. An international team of astrophysicists led by John Webb, of the University of New South Wales, Sydney, had analysed light coming from the edge of the universe. They concluded that the value of a constant of Nature, familiarly known as *alpha* ($\alpha$), seems to have been smaller 10 billion years ago.

Why the surprise? $\alpha$ is of interest because it embraces three fundamental constants of physics: the charge of the electron; the speed of light in a vacuum (which, according to Einstein's special theory of relativity, is the ultimate limit of speed in the universe); and Planck's constant, which governs the world of the microscopic, and the fascinating field of quantum mechanics. These three fundamental constants are considered to be immutable entities of physics. Consequently, as $\alpha$ is a combination of these quantities, it cannot vary.

Did this observation indicate a seismic shift in contemporary physics, or was it a nonsensical storm in a teacup? Let us imagine that we belong to an international commission of experts, charged with an inquiry into the solidity of modern physics. How do we evaluate the theoretical implications of this communication reporting a possible variation of $\alpha$?

## SEARCHLIGHTS INTO THE PAST

*Alpha* ($\alpha$), known as the 'fine-structure constant', characterises the interactions between matter and light. It has been very accurately measured in the laboratory. It is indeed the most precisely measured of all physical constants with a value of 0.007297352568, or, otherwise expressed, 1/137.03599976. It is best memorised in the form $\sim 1/137$.

Why did the Sydney physicists seek to evaluate $\alpha$ in the distant Universe, when it had already been determined with such accuracy in the laboratory? Presumably, as the laboratory measurements have all been made in recent decades, the Australian researchers would want to know if $\alpha$ had maintained the same value through time. Of course, looking deeply into space is tantamount to looking far back in time. This is a consequence of the finiteness of the speed of light. If, for example, we observe a solar flare by optical means, we see the event with a slight time delay, since the light has taken just over 8 minutes to reach us from the Sun. If we then turn a telescope towards Proxima Centauri – the nearest star to the Sun – we see it as it was a little over 4 years ago. The more distant the objects of our scrutiny, the older the phenomena we observe. Thus, to track the past value of $\alpha$, we need to look at the most distant regions of the universe, searching for phenomena that may be more susceptible to any change in the value of the constant. These phenomena occur in relatively 'empty' regions, but the universe is not as empty as it seems because immense clouds of gas and dust lie between the galaxies. As these clouds do not emit light, they are difficult to observe directly; in fact, atoms within the clouds absorb some of the light traversing them – a phenomenon that depends on the constant $\alpha$. This light should therefore give some insight into the value of $\alpha$.

To go as far back in time as we can, we use the light emitted by the most distant observable objects, and quasars have been found to be particularly good candidates. These galactic nuclei are extremely energetic and began to shine about 10 billion years ago, at a time when the universe was young. Today, we can observe these powerful 'searchlights' towards the edge of the observable universe. Their light pervades the remainder of the cosmos, especially the clouds of gas and dust mentioned above. On its journey, light from a quasar can interact with atoms in these clouds and leave traces of its passing. By collecting this evidential light, physicists can deduce the value of $\alpha$ at the time of these interactions. They examine clouds at different distances, from about 10 to 12 billion light-years away, thereby determining the value of the fine-structure constant at various epochs in the distant past.

This is precisely what John Webb and his colleagues did in 1999, and have done several times since. Their conclusion is that, contrary to all expectations, the value of $\alpha$ was smaller by 1 part in 100 000, some 12 billion years ago. A hundred-thousandth part? That doesn't seem much! Irrespective of how small this variation might be, it could nevertheless throw a spanner into the well-oiled works of modern physics.

## INSPIRED – OR CRANKY?

An inconstant constant? What a preposterous idea! However, studying this claim more closely, we hear echoes of a debate, more than 70 years old, about this very subject: the constancy of fundamental constants. The background to this debate involves the properties of gravity and possible alternative theories to Einstein's general relativity. Indeed, Webb and his team are not the first to question the constancy of fundamental constants. There have been some very passionate, and famous, debates among researchers. The first to air such a possibility was Paul

Figure 1 Paul Adrien Maurice Dirac (1902–1984). (AIP, Emilio Segré, Visual archives.)

Adrien Maurice Dirac, one of the founders of quantum mechanics. Dirac was passionate about the mathematical beauty of physical laws (Figure 1). Considered to possess a superior and rational mind by his peers, Dirac received the Nobel Prize in 1933, at the age of 31, for his prediction of the existence of the anti-particle of the electron, known as the positron. Dirac was never afraid to publish hypotheses that some thought to be 'cranky'. He was certainly no idle dreamer, and put his ideas rigorously to the test through theoretical reasoning or tightly controlled experimentation.

In the 1930s, Dirac began the quest for a 'fundamental theory' embracing microphysics and the description of the cosmos. He followed the trail left by one of his former professors at Cambridge University, the astrophysicist Arthur Eddington, who was famous for having carried out, among other things, the first verification of one of the predictions of Einstein's general theory of relativity. Both Dirac and Eddington based their approaches on the question of the values of fundamental constants, such as the gravitational constant, the speed of light, and the charges and masses of the electron and the proton. No known theory predicted their values: it was necessary to measure them.

Conscious of this problem, Dirac tried to reveal a hidden structure by manipulating the fundamental constants, and used some algebraic combinations to see what values would emerge. This numerological 'game' led Dirac to incredibly small and large numbers which never ceased to amaze him. He compared, for example, two fundamental forces acting between a proton and an electron: on the one hand, the electric force (depending on the square of the electric charge), and on the other, the gravitational force (depending on the product of the masses of the two particles, and on the gravitational constant). The ratio of the strengths of these two forces was 'approximately' $10^{39}$: it seems of little relevance to be so precise – a 1 followed by 39 zeros! A thousand billion billion billion billion!

Such enormous figures shocked Dirac's sense of harmony. How could one conceive of such disparate quantities being unified in a 'fundamental theory'? Dirac then investigated the large numbers that emerge when we try to describe the universe in terms of atomic physics. In particular, he had the idea of comparing the age of the universe with the period of the orbit of the electron around the proton in the simplest of all atoms: hydrogen. The relationship between these two periods was... $10^{39}$! Well, well! A pure numerical coincidence, or a hint of the existence of some undiscovered law? Dirac could not believe that mere numerical coincidence was working here. He preferred to see in the relationship evidence of a more fundamental physical theory, which he called the 'large numbers hypothesis'.

> "... all very large dimensionless numbers which can be constructed from the important natural constants of cosmology and atomic theory are connected by simple mathematical relations involving coefficients of the order of magnitude unity. ..."

Let us see what Dirac deduced from his large numbers hypothesis. He assumed that the two large numbers in question must always be equal, and not only in the present: the ratio of the electric force to the gravitational force between the proton and the electron would be proportional to the age of the universe. In theory, the first term (the ratio of the forces) remains constant with time, while the second (the age of the universe) increases with time. If, however, we allow the first term to vary with the second term, its large value is explained: it is nothing but a consequence of the great age of our universe. For this to be the case, one or more of the fundamental constants involved in the relationship between the electric and gravitational forces must change with time. Dirac assumed that the gravitational constant varied in inverse proportion to the age of the universe, thereby diminishing through time, while the values of all the other constants remained fixed.

This hypothesis, based as it was on purely numerological arguments, came as a surprise to more than one of Dirac's colleagues: from him, they expected an implacable rationalism. There were those who whispered that Dirac's recent honeymoon trip had addled his mind, as he was married in 1937, and soon afterwards published his 'large numbers hypothesis' in the famous British scientific review *Nature*. Although few took this work seriously, the seed had nevertheless been planted: the possibility that a fundamental constant could vary was growing.

## THE JOYS OF OXYMORON

"Varying constants"? We are reminded of the famous line from *Le Cid* by Corneille: "*Cette obscure clarté qui tombe des étoiles*" ("That obscure brightness that falls from the stars"). Had physicists nothing better to do than dabble with the joys of oxymoron?

In reality, the question of the 'constancy' of fundamental constants entails far more than just amusement for idle physicists. As proof of this, we can show that some of Dirac's thoughts and motivations were rooted in the bedrock of physics, and have relevance to this day. We shall select three illustrative examples.

Firstly, Dirac applied to the universe in general a relationship between two different values, validated only in the present and locally: i.e. on Earth and in its vicinity. He suggested implicitly that the laws of physics are valid everywhere and at all times. If the relationship reflected some law of physics, it had to obey the same principle of *universality*. The universality of the laws of physics is a basic tenet of current cosmology: no place or point in the universe possesses properties that are specific only to itself. In particular, the universality and immutability of constants underpin the modern description of gravity as understood through Einstein's general relativity. Consequently, the verification of those properties of constants tests the laws of today's physics.

Secondly, Dirac was motivated by the desire to bring together the various descriptions of the physical world, notably that of the microcosm of atoms and that of the macroscopic universe. *Unification* is one of the driving forces in physics. Newton was among the first to find such a unification, describing with a single law how bodies fell to Earth and how planets orbit within the solar system. Today, three of the four fundamental interactions (the electromagnetic, and the weak and strong nuclear forces) have been unified within one physical framework, while gravity, described through general relativity, remains aloof. In passing, we note that, among the most promising theories aiming at this unification, string theory could accommodate the idea of varying constants.

The problem of unification is connected with Dirac's third point: the gravitational force experienced by two charged particles is minuscule compared with the electrical force acting between them. This disparity, known technically as the 'hierarchy problem', is even now still not understood, and remains an obstacle on the path towards the desired unification.

To these three points, arising directly from Dirac's reflections, we can add a fourth and even more fundamental one: to inquire into the constancy of constants is to question the very laws in which they are involved. It is in fact to question the whole of physics, since all its equations bring in fundamental constants!

How did constants arise? What is their role in modern physics? How many of them are there? Should they all be seen as having equal standing? What would become of the laws of physics if one of the constants were shown to vary? A 'true' constant must not vary, either in time or space. Why is this property so important? The questions surrounding constants are many, and our 'commission of inquiry' will consider them all.

## 'CONSTANTOLOGY'

For the best part of 200 years, the values of fundamental constants have been measured in the laboratory. If any one of them had varied significantly during that time, we might think, in our naivety, that the variation would have been detected by now! The reproducibility of laboratory experiments – the pillar of all scientific method – implies that if there has been any variation, it is very slight and is within the limits of measurability. Experimental verification of the 'constancy of constants' therefore requires highly accurate metrological techniques, at the limit of current capabilities. The study of fundamental constants exercises growing numbers of researchers, both experimental and theoretical. It can even be seen as a discipline in its own right, which we might call 'constantology': the subject of this book.

Constantology is a vast undertaking, embracing as it does all the disciplines of physics: in just a few lines, we have already touched on atomic physics, with its interactions between atoms and light; astrophysics, with its quasars and intergalactic clouds; cosmology, involving the birth and evolution of the universe; and gravitational theories, general relativity and string theory. Throughout our inquiry, fundamental constants will take us on a journey through both time and space, from the beginning of the universe to the present day, and from the laboratory out to the far reaches of the cosmos. We shall also, more particularly, navigate the universe of physical theories.

The question before us might be expressed thus: Does the variation of a fundamental constant threaten to topple the edifice of current theories? Our inquiry will begin to answer this question by first examining that edifice. After inspecting the foundations, we shall establish those constants that most solidly support the structure. We shall then identify the 'shakier' supports, i.e. the constants or parameters whose possible variation would lead to repercussions. At the same time, we shall take a close look at the world of one of these parameters, the famous $\alpha$. Lastly, we shall come back to those theoretical clues to an imminent revolution in physics.

Therefore, with notebooks at the ready, magnifying glasses in hand, and CCD and digital cameras prepared, let us begin our inquiry.

# 1

# Foundations: universality in physics

Before we begin, let us plan out our mission. Do we really know what the subject of our inquiry is? We know it is related to 'fundamental constants' or 'universal constants', but can we define these terms precisely?

Since primary school we have been in the habit of looking in a dictionary if any doubt arises about the meaning of a term. The 2004 edition of the French *Larousse* dictionary defines a constant in these terms:

> "**Constant**: in physics, the numerical value of *certain* quantities characterising a body" (our italics).

You will note that we have emphasised the word 'certain'. What quantities are we dealing with? What distinguishes these quantities from others? The dictionary offers no answer to these questions, but the definition continues:

> "A particular quantity of fixed value (the mass and charge of the electron, Planck's constant, *for example*), which plays a *central role* in physical theories."

The dictionary offers a short list, and follows with the terse 'for example'. What else might be in that list? How many fundamental constants are there, and what is this central role they are supposed to play in physical theories? We are hardly any further forward in our inquiry.

From another French dictionary, the *Petit Robert*, we read:

> "**Constant**: a quantity that maintains the same value; a number independent of variables."

Under 'physics', the dictionary defines the expression 'universal physical constant' as a "quantity which, measured within a coherent system of units, is invariable". But how do we know, *a priori*, that a quantity is invariable? Can we unambiguously distinguish a constant from a variable?

Still immersed in our dictionaries, let us now look at the word 'fundamental':

> "**Fundamental**: acting as a foundation. Serving as a basis; an essential and determining characteristic."

Here we find the essential characteristic of the aforementioned laws: a 'fundamental constant' seems to be a foundation of a physical theory.

Obviously, the dictionary offers only a superficial definition of a fundamental constant. However, it does underline the essential part that fundamental constants play in the theories of physics. Our commission of inquiry must therefore begin its inspection among the foundations of the physical edifice, wherein lie the founding principles of physics; we shall doubtless find that fundamental constants are some of their natural consequences. Switch on your miner's lamps: we are descending into the basement.

## UNIVERSAL TRUTHS

The first stage of our inquiry into the foundations of physics takes us to Poland, in the 1510s, and in particular to Frauenburg (Frombork), a sumptuous port on the banks of the Vistula. Here, Nicolaus Copernicus (Figure 1.1), a humanist priest well versed in mathematics, astronomy, medicine and theology, had an observation tower constructed from which he intended to study the Sun, the Moon, the planets and the stars. In his efforts to describe the motions of the celestial bodies, Copernicus tried to improve upon the system that was elaborated by Claudius Ptolemaeus (Ptolemy of Alexandria), in the second century AD, and based upon the teachings of Aristotle. The Earth was considered to be motionless at the centre of the heavens while the other celestial objects revolved around it on circular paths known as epicycles. Ptolemy's system

Figure 1.1 Nicolaus Copernicus (1473–1543). (Courtesy of the University of Texas.)

described the motions of the planets faithfully enough, including their whimsical habit of turning back occasionally along their own paths.

However, this geocentric model did not satisfy Copernicus. His intuition told him that things were arranged differently, and he held that the Sun – and not the Earth as Aristotle had claimed – lay in the centre of the universe, Copernicus also thought that the Earth must resemble the other planets, which he had long studied, and believed it to be spherical, rotating in one day and revolving around the Sun in one year. It seemed to him more logical to ascribe to the Earth "the motion appropriate by nature to its form, rather than attribute a movement to the entire universe, whose limit is unknown and unknowable". A serious crack had appeared in the idea that the Earth is an exceptional, or even a unique, body!

With his heliocentric model, Copernicus replaced Ptolemy's complicated system by a general scheme supported by a few fundamental principles, one of which was that the Earth is a planet like the others. More importantly, Copernicus removed the Earth from its hitherto central and privileged position.

This concept differed radically from that of Aristotle, which had prevailed until then. It was, however, Aristotle who had laid down the scientific method, basing it upon principles: "Wisdom is knowledge about certain principles." He even distinguished between terrestrial and celestial laws, but such a distinction loses all meaning in the Copernican system, where our position as observers of the universe is no longer unique. It had become possible to establish the properties of the universe independently of the position of the observer, and it was from this great conceptual revolution that physical science evolved into its modern version. The science of physics then tried to establish universal laws that were valid everywhere – what we call today the 'Copernican principle'. This principle has sometimes been extended to the validity of laws in time as well as in space: that is, laws established today were identical in the past and will be in the future. Physical science had become a quest for laws that were both universal and *eternal*.

## HARMONIOUS ELLIPSES

Copernicus' hypothesis met with rejection from theologians both Catholic and Protestant, and provoked incredulity among astronomers. Why should we abandon the Ptolemaic model, which has served us so well? Why should we burden ourselves with a hypothesis that would bring nothing but deep philosophical confusion? Nevertheless, the revolution had begun. By the middle of the sixteenth century, it had reached Benatek Castle in Prague, where Copernicus' heliocentric model was carefully scrutinised. The castle, which was also an observatory, belonged to a Danish nobleman-astronomer named Tycho Brahe (Figure 1.2).

A man of his time, Brahe worked out a hybrid model drawing on both the Ptolemaic and Copernican systems: the Earth was motionless at the centre of the universe (at that time an accepted fact), the Moon and the Sun moved around

Figure 1.2 Tycho Brahe (1546–1601) in his observatory in Prague. (Bibliothèque Nationale de France.)

Figure 1.3 Johannes Kepler (1571–1630). (Courtesy of the University of Texas.)

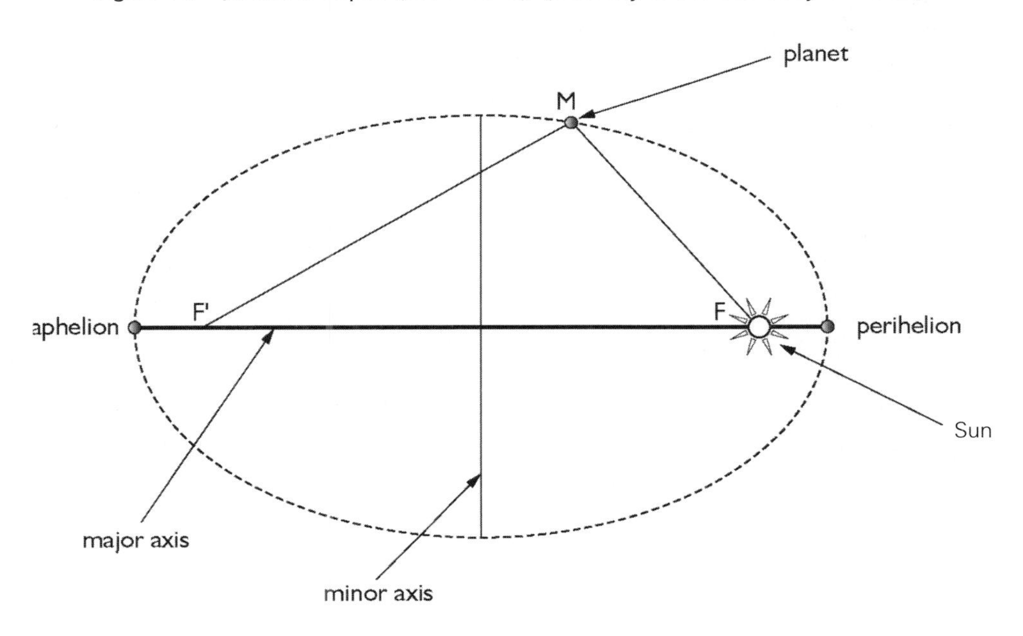


Figure 1.4 An ellipse and its two foci, F′ and F.

the Earth, and the other planets moved around the Sun. Brahe thought that he could refine this model using his own observations, the most accurate of the day. A young German, Johannes Kepler (Figure 1.3), who assisted Brahe in this, was less sure of his master's hybrid model, preferring that of Copernicus. He worked on the problem of the planet Mars, whose orbit seemed to depart from the perfect and divine circle. He showed that the shape of Mars' orbit is in fact oval rather than circular – an ellipse, which has not one but two foci, one of which is the Sun (Figure 1.4).

The collaboration between Kepler and Brahe was both brief and stormy. After Brahe died of a ruptured bladder while dining, Kepler inherited the planetary data that had been kept from him for so long. After eight years of painstaking work, Kepler deduced from these data the three laws that now bear his name. They describe quantitatively the motion of every planet. The first of these laws states that all planets, including the Earth, move, like Mars, in elliptical orbits. The Sun is at one of the foci of the ellipse. Kepler's second law, the 'law of areas' (Figure 1.5), states that the line between the Sun and a planet sweeps out equal areas in equal times in the course of its orbit. The third ('harmonic') law tells us that the square of the period of revolution of a planet is proportional to the cube of the semi-major axis of the ellipse it describes.

Kepler's three laws offered a complete alternative to the Ptolemaic model. Only three laws, founded upon geometrical considerations, described the motion of all planets and made it possible to predict their positions. The problem now was to ensure their general acceptance.

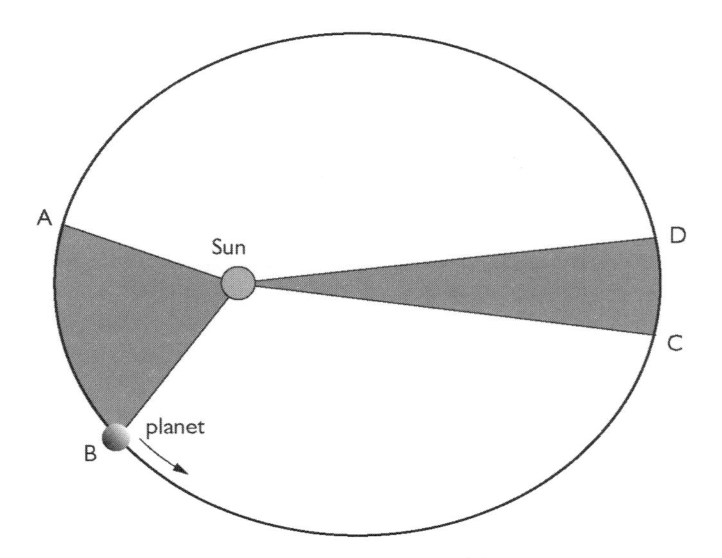


Figure 1.5 Kepler's second law: a planet sweeps out equal areas in equal times, so that travelling from A to B takes the same time as travelling from C to D.

Figure 1.6 Galileo Galilei (1564–1642). (Courtesy of the University of Texas.)

## INERTIA AND RELATIVITY

Kepler felt isolated in his beliefs, and, around the end of the sixteenth century, he canvassed the support of an exceptional Italian scientist, Galileo Galilei (Figure 1.6), for the ideas of Copernicus. Galileo did not immediately side with Kepler, for he was immersed in the mysteries of motion itself. He was working on the elaboration of his theory of mechanics – a subject he taught at the prestigious University of Padua. Like Copernicus before him, Galileo was not comfortable with the Aristotelian doctrine for the simple reason that it did not stand up to experimental testing. Experimentation is, after the Copernican principle, the second driving force of modern physics. Whereas Aristotle had collated simple observations and common sense, Galileo, like any modern scientist, 'chipped away' at Nature in an attempt to discover its laws. The Italian master studied pendulums, rolled balls down inclined planes, and dropped bodies into media of different viscosities. He also created thought experiments, placing himself mentally in experimental conditions unachievable in reality, in order to distil the essence of certain physical phenomena, and was the first to use this intellectual process that is now beloved of modern physicists.

Galileo finally reached an understanding of the nature of motion. A force applied to an object causes it to move, but when this force ceases to act, does the object cease to move, as Aristotle had claimed? Galileo was not so sure, since a ball rolling along a smooth and perfectly horizontal table continues to roll for a time, even although no force is in evidence and the duration of this movement increases in inverse proportion to the amount of friction offered by the table. Galileo therefore extrapolated his observations, imagining a world with neither friction nor resistance, a void. His rigorous observations and capacity for abstract thought led him to announce his principle of inertia: in a vacuum, a moving body, even though no force is acting upon it, will continue to move indefinitely in a straight line and at a constant velocity.

It was Galileo who conferred a particular status to inertial, rectilinear and uniform motion, affirming that "movement is as nothing". Let us imagine, as Galileo did, that we are in a cabin on a boat, moored at the quayside in calm waters. We observe the flight of a butterfly, which is in the cabin with us, and the fall of a drop of water from the ceiling of the cabin into a bucket directly below it. The butterfly flies randomly in various directions, and the drop of water falls directly into the bucket. When the boat is sailing at a constant speed – still on a calm sea – we observe the same things to be true. The butterfly does not fly preferentially to the fore or to the rear of the cabin, and the water still falls vertically. There is no way of telling whether the boat is moving or not: while the vessel remains rectilinear and uniform, its motion is "as nothing". It is only when the boat accelerates, decelerates, turns or pitches, causing the speed to change, is any difference is perceptible.

Galileo therefore showed that it is impossible to distinguish immobility from uniform rectilinear motion. Modern travellers will understand this: we have all looked through the window of a train at another train passing by, and have not immediately known which of the trains is actually moving. As the notion of movement and immobility is 'relative', it has become known as the 'principle of Galilean relativity'. Three centuries later, Einstein would incorporate that principle into his special theory of relativity. But let us not rush too far ahead; suffice it to say that this principle constitutes a robust pillar of our edifice.

It should be noted that this is also an extension of the Copernican principle: the laws of motion must be identical for two observers moving relative to each other, uniformly and in a straight line. In order to pass from one viewpoint to the other, we need only add to the velocity of one observer the relative velocity of the other. For example, if we walk along the corridor of a train travelling at 60 kilometres an hour, our speed relative to the neighbouring compartment is 5 kilometres an hour, but it is 65 kilometres an hour relative to the rails beneath. These so-called 'Galilean transformations' from one frame of reference to another, have seemed obvious, and verifiable through experiment, for centuries.

## THE FALL... OF ACCEPTED IDEAS

We stay in the time of Galileo, in order to see the consequences that flowed from his principles. At the beginning of the seventeenth century, the Italian scientist was revolutionising astronomy. He improved the design of optical lenses and telescopes, and, pointing his instrument skywards, he saw the heavens magnified and the surface of the Moon crowded with mountains and craters. He also discovered the nature of the Milky Way, which consisted of multitudes of stars, and the universe now seemed to be much vaster than had ever been imagined. All these marvellous discoveries were presented in 1610 in Galileo's *Starry Messenger*, which earned him wide renown, and the time had come to profit from his fame. He published his thoughts on Aristotle's ideas, and confirmed the Copernican theories, with special reference to observations he had made of the phases of Venus.

Galileo's vision, judged contrary to the established teaching of both the church and Aristotle, was condemned by the Inquisition in 1613. Such a condemnation gave pause for thought: in 1600, the free-thinker Giordano Bruno had been burned at the stake for having affirmed the existence in the universe of an infinity of worlds like those of the solar system. However, Galileo cannot be compared to the turbulent and 'heretic' Bruno. He knew that his affirmations had a solid foundation, and he would not be so easily silenced.

Imprisoned in 1632 as a result of having published his *Dialogue Concerning the Two Chief World Systems*, Galileo worked on a new paper entitled *Dialogues Concerning Two New Sciences*, which was published in 1638. In this paper he revised and refined his work on mechanics, and in particular the science of falling bodies. Galileo's perfected method would give physics its first *lettres de noblesse*, and set the code of practice for physicists. Experimentation would now rely on reproducible results; theory would furnish predictions; and the interplay between experiment and theory would lead to the establishment of the most general of laws. Galileo had previously proved that the speed of a falling body increases with time. He was now convinced that, in a vacuum, this increase (acceleration) was the same for all bodies, from leaden balls to feathers, independent of their mass and chemical composition. In other words, *all* bodies fall in the same way in a vacuum. This principle of the equivalence of falling bodies, also known as the universality of free fall, would be taken up again by Newton, and later by Einstein. It is another pillar of physics, and our commission of inquiry will attempt to test its solidity.

## 'GRAVE' REVELATIONS: GRAVITY

During the first decades of the seventeenth century, Kepler described the orbits of the planets around the Sun, and Galileo provided a model for objects falling to Earth. In the second half of that century there arose a giant of physics who would make the connection between these two phenomena. Isaac Newton gave physical science nothing less than its first mathematical formulation and the first unification of its concepts.

The famous legend has it that an apple fell from a tree beneath which Newton was dozing, inspiring his amazing idea of a single unique force that would be responsible for the fall of apples *and* the motion of the planets around the Sun.

Even if we are somewhat dubious about the idea of Newton having regular *siestas*, we can still imagine him working in his rural retreat on his theory of gravitation. In 1665, an epidemic of bubonic plague spread from London and threatened Cambridge, where Newton was completing his studies. The closure of Trinity College obliged him to return for 14 months to his family home in Lincolnshire, in eastern England. This enforced leave proved so fruitful that Newton later described it as *'mirabilis'*. The young physicist was interested in philosophy, mathematics, and the new ideas in mechanics and astronomy arising from the work of Copernicus, Kepler and Galileo. In homage to his predecessors in the field of physics, Newton was to write "If I have seen further, it is by standing on the shoulders of giants".

Taking up where Galileo left off, Newton made the principle of inertia his own. Any body not subject to the influence or action of another will move in a straight line at a constant velocity, or be at rest. Universality is ever-present – the law deals with *any* body. Newton also noted that if a force causes an object to move, it does so by modifying its velocity: the acceleration of that object (the rate of change in velocity) is proportional to the force applied.

When, at the end of this productive year, Newton assumed the chair of mathematics at Trinity, a debate was current among scientists concerning the mathematics of the force of attraction between the Sun and the Earth. The Astronomer Royal, Edmond Halley, put a question to Newton: What would be the shape of an orbit if this force varied inversely with the square of the distance between the two bodies? The great physicist immediately replied... an ellipse. For Newton had already solved the problem, using Kepler's laws. He had deduced from the 'law of areas' that the attractive force between the Sun and a planet acts along the line between their centres. From the 'harmonic law', he inferred that this force is inversely proportional to the square of the distance between the Sun and the planet: double the distance, and the strength of the force is one-fourth of the original; treble it, and it falls to one-ninth. Unfortunately, Newton was unable to find his calculations to show to Halley!

Halley was insistent, and Newton eventually managed not only to reconstitute his proofs, but also to produce a summary of all his work on mechanics: *Philosophiae Naturalis Principia Mathematica* (*Mathematical Principles of Natural Philosophy*), published at Halley's expense, in three volumes, in 1687. The first volume deals with his three laws of motion, the second covers fluid mechanics, and the third offers a physical description of the whole universe, including the demonstration requested by Halley. Working from the three laws of motion, Newton establishes the existence of "a gravitational effect common to all bodies, proportional to the total quantity of matter contained within them (their mass)". Two bodies attract each other with a force that is proportional to the product of their masses and inversely proportional to the square of their distance apart. Notice that Newton here states that gravity is "common to *all* bodies". Which bodies has he in mind? The apple falling from the tree, drawn down by the considerable mass of our planet; the Moon, held in its elliptical orbit by the Earth; the Earth and its fellow planets, and even comets, attracted towards the

Sun. In short, the law of gravity applies to all masses within the universe, on Earth as well as in the heavens. It is indeed a universal thing, even more so than Galileo's law of falling objects, which applies only to things falling to Earth, and Kepler's laws, which deal only with celestial objects. Newton's law unifies the motions of the planets and of falling bodies, attributing a common cause: that of universal gravitation.

This economy of explanation was clearly announced by Newton when he laid down, at the beginning of the third volume of the *Principia*, rules to be followed in the study of the physical world. The first two are:

I. We are to admit no more causes of natural things than are both true and sufficient to explain what is observed.
II. Therefore to the same natural effects we must, as far as possible, assign the same causes.

The same criteria of universality and unified explanation apply in modern-day physics.

## MEASURE IN ALL THINGS

Learned men like Copernicus, Kepler, Galileo and Newton added their ingredients to the cauldron of universality in which the fundamental constants would simmer. How did these constants come to be? To answer this question, we must look to another aspect of physics: measurement.

Measurement involves the comparison of objects, though the physicist might call them *systems*. For example, in his work on falling bodies, Galileo compared the velocities attained by two bodies after a certain time had elapsed: the ratio of the velocities is equal to that of the times during which they have fallen. Newton reasoned in the same way: he was examining the relationship between physical quantities of the same kind. This method prevailed for a long time, but in order to be able to compare experiments conducted in different places, physicists began to adopt another approach. They defined an ideal reference system – a system of units – that would be the same for all physicists, would always be repeatable, and could be used when comparing the issues being studied.

An example will serve to illustrate this approach, and to introduce the idea of units. We can describe a pole in terms of both its length and mass. We measure its length by comparing it with a ruler, chosen as a reference object of a standard length, which we call a *metre*. So, we state that the pole is '3 metres long', if the metre rule must be laid along it three times to establish the length (Figure 1.7).

Similarly, the mass of the pole is obtained by comparing it, by means of a balance, to a reference mass: let us call this reference the *kilogramme*. The metre and the kilogramme are the *units* of measurement of the pole. The results of the measurements, i.e. the operations of measuring, take the form of two pure numbers: the ratio of the length of the pole to that of the reference metre rule, and the ratio of its mass to the reference kilogramme. Suppose that we obtain, as

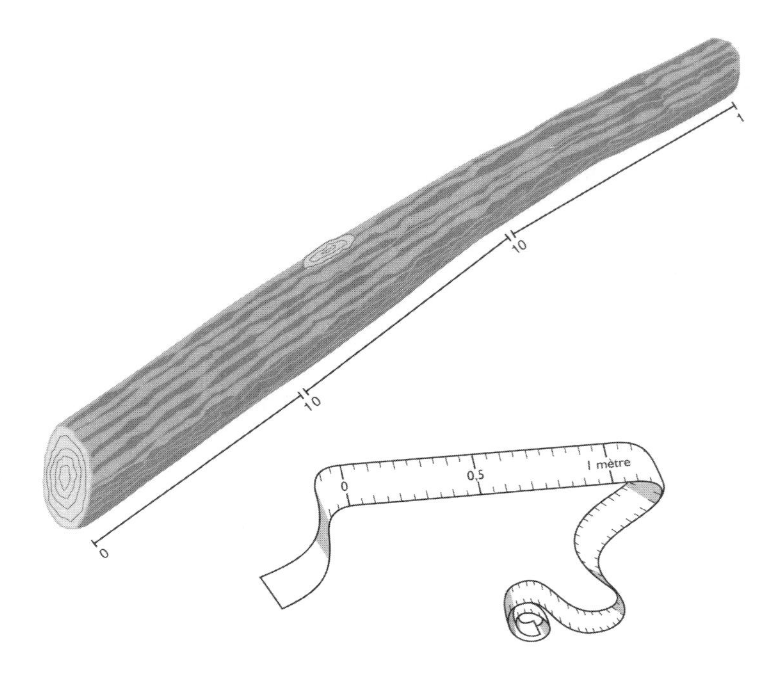


Figure 1.7 Measuring a pole 3 metres long: the reference metre, laid along it three times, gives its length.

results, 3 and 2. This means that our pole is 3 metres long and weighs 2 kilogrammes. If we communicate the measurements to colleagues, they will be able to fashion an identical pole. The units of measurement will serve as references for comparison during the operation. This will be possible if all scientists agree on the standard mass and length to use, using the same definition of the metre and the kilogramme.

Thanks to such units, physicists can study systems in an 'absolute' frame of reference: all agree on how to compare a quantity with a system of reference. We see here an evolution in the way in which observed phenomena are described. Looking at falling bodies, the statement:

> "the ratio of the velocities of the two bodies is equal to that of the times during which they have fallen"

becomes:

> "the velocity of any falling body is proportional to the time during which it has fallen".

We agree that the velocity of the body should be compared with a reference velocity, and that duration of the fall be calculated on the basis of a standard unit of time.

What do we mean by 'proportional'? It is not easy to see how we can make a velocity equal to a time. Since the nature of each of the two quantities is different, they can only be related to each other by bringing into play another quantity, a constant of proportionality which takes account of this difference in natures. In our example of free fall, the constant in question is $g$, representing the rate of change in velocity as time passes: acceleration. So a 'constant' such as $g$ acts within a fundamental law.

This new approach was not adopted overnight. The transition seems to have occurred around the end of the eighteenth century, a little earlier in France than in England. For example, in 1750, French astronomer Alexis Claude Clairaut was already using it, while English physicist and geologist John Michell was still, in 1784, relying on Newton's beloved 'ratios of quantities'. The success of the system of units and of constants reflects their ease of use, notably when physical phenomena other than motion, such as electricity and magnetism, are involved.

## THE VALUE OF A CONSTANT

Thus, in their modern form, the laws of physics involve two types of terms: *variables*, representing characteristics of the system studied (e.g. velocity, duration of fall), and *constants*, which allow us to compare different quantities, for example the constant $g$, in the law of falling bodies. Variables are implicitly compared with reference systems, which means that the 'user' of the physical law must specify the nature of the quantities involved: velocity, duration, acceleration, etc.

It is not possible to compare, add or subtract quantities if they are not of the same kind (or, in the terminology of physicists, of the same *dimensions*). Similarly, in an equation describing a physical law, both sides of the equation must represent the same quantity/dimensions.

This simple requirement enables us to identify the nature of a constant. In the law of falling bodies, velocity $v$ equals the time $t$ multiplied by the constant of proportionality $g$. This can be summarised using the mathematical equation $v = gt$. The velocity is calculated by dividing the distance travelled by the time taken to travel it. It therefore has the dimensions of a length divided by a time ($L/T$, $L$ and $T$ being the conventional notations for the dimensions of length and time, respectively). In order that the second part of the equation should also possess the dimensions of a velocity, the constant $g$ must have the dimensions of a length divided by the square of a time: $L/T^2$, representing an acceleration.

Other constants have less intuitive dimensions than $g$; for example, Newton's gravitational constant. According to Newton, the force of gravity (which we call $F$) between two bodies is proportional to the product of their masses ($m_1$ and $m_2$), and inversely proportional to the square of the distance ($r$) between them. The modern equation expressing this idea is written $F = Gm_1m_2/r^2$, where $G$ is the gravitational constant. What are the dimensions of $G$?

Staying with Newton, we express a force as equal to the product of a mass $m$ and an acceleration $a$: so $F = ma$. The force therefore has the dimensions of a mass multiplied by a length, divided by the square of a time ($ML/T^2$, where $M$ represents a dimension of mass). That satisfies the first part of the equation. In the second part, we have a mass multiplied by another mass and the inverse of the square of a distance. In order that the two parts of the equation may balance, $G$ must possess the dimensions of the cube of a length divided by mass and by the square of a time ($L^3/MT^2$).

We can now ascribe a value for each of the quantities present by specifying a system of units. If we use the units of the *Système International* (SI), the masses are expressed in kilogrammes, the lengths in metres and the times in seconds. The constant $G$, measured in cubic metres per kilogramme per second squared, becomes $6.67 \times 10^{-11}$. This is a very small, near-zero value. The negative power of 10 here ($10^{-11}$) means that the first significant figure (6) is in the eleventh position after the decimal point.

Now, if we adopt other units, this value will be different. Let us, for example, use millimetres instead of metres, tonnes instead of kilogrammes, and minutes instead of seconds. We then obtain a (perhaps more intuitive) value of 185. So, the numerical value of a constant assumes meaning only if we specify the system of units used to express it.

## A FRUITFUL EXPEDITION

What have we uncovered among the foundations of our edifice? Firstly, the principles upon which science rests and, secondly, the first inkling of its fundamental or universal constants. Indeed, it is physical science as it has been conceived that causes fundamental constants to conserve their values, whatever the system being studied, wherever in space that system might lie, and at whatever epoch it is studied.

Copernicus is often credited with the idea of the universality of the laws of Nature: although he never used the term himself, it is a way of paying homage to his forward-thinking work. Since we do not occupy a privileged place in the universe, the laws of physics we establish for our vicinity must also apply to the rest of that universe.

To handle relationships between different dimensions, the laws of physics apply fundamental constants, which naturally inherit their properties of universality. The constants, together with the principles from which they arose, serve as pillars supporting the edifice of physics. Our commission of inquiry will now turn its attention to these items.

# 2

# Fundamental constants: pillars of the edifice

Leaving behind the foundations, our commission of experts now begins its inventory of some of the pillars underpinning the edifice of current science: the fundamental constants. In the course of our explorations, we have already encountered two constants: the acceleration due to gravity, $g$, and the gravitational constant, $G$. They appeared in modern formulations concerning falling bodies and universal gravity introduced, respectively, by Galileo and Newton. They are called constants, as opposed to the variables, such as velocity, time (duration) and position, which they link. Is this their only characteristic? Do they all have equal importance, or are some more essential than others?

## AN INVENTORY OF CONSTANTS

On the basis of the two constants, $g$ and $G$, already encountered, we can propose a general definition of the object of our study: a "constant" is any non-determined parameter which appears in the formulation of physical laws. This is a first step in our investigation. In order to take further steps across the *terra incognita* of fundamental constants, we find the method adopted by naturalists to be the best suited: when they uncover some new concept, they make an inventory of everything in that category. They produce lists and catalogues, to reveal fundamental characteristics.

Where can we find physical constants? It might be a good idea to start by opening a physics textbook! A list of formulae worthy of the name will show the values of constants, to assist both the junior and the experienced physicist in their calculations. What can be found, for example, in a first-year university textbook, with constants arranged according to the disciplines of physics?

- Under mechanics, we find the acceleration due to gravity (at the latitude of Paris) $g$, and the gravitational constant $G$, which we have already met.
- Under thermodynamics (the science of heat and machines involving heat), we find the triple point of water. This is the temperature at which the three phases of water (ice, liquid and vapour) are in equilibrium. The triple point is the basis of the definition of the Celsius temperature scale. There is also the

ideal gas constant, denoted by the symbol $R$, bringing together pressure, volume, temperature and the number of microscopic constituents of a tenuous (or *ideal*) gas. Boltzmann's constant, $k$, relates the temperature of a gas to the velocity of its constituent particles.

- Under electromagnetism, we find the inevitable constant $c$, familiarly known as 'the speed of light': it represents the speed of the propagation of any electromagnetic wave through a vacuum. It is a constant with a long history, which we shall examine later in some detail. The second indispensable constant in electromagnetism is $e$, the elementary charge. It represents, as an absolute value, the smallest possible electric charge carried by a free particle: for example, the electron carries the charge $-e$, and the proton the charge $+e$.
- Under quantum mechanics, the discipline which describes the microscopic, we find a constant of paramount importance: Planck's constant, $h$. We shall also recount the history of $h$. Other useful constants include the electron rest mass $m_e$ and the proton rest mass $m_p$. There is also the fine-structure constant $\alpha$, which we have already mentioned, and which is currently furrowing the brows of many an experimental physicist.

## THE FLAVOUR OF THE TIMES

However, had we consulted a different physics book, we might have found a different list. Certainly no appendix gathering the constants, etc. at the end of a book published in the nineteenth century will list Planck's constant or the fine-structure constant, which are both twentieth-century creations. We would, however, find plenty of constants from the disciplines of thermodynamics and mechanics, which then ruled the roost.

Any given list of fundamental constants therefore reflects the *ensemble* of the laws of Nature as understood at the time of its compilation. Constants in the list can be determined only through experiment. As the nineteenth century drew to a close, scientists believed that they were about to reach the summit of all knowledge. The equations of classical mechanics, thermodynamics and electromagnetism seemed to be all that was needed to describe the world. Indeed, the brilliant Scottish physicist, James Clerk Maxwell (Figure 2.1) stated in his inaugural address at the University of Cambridge in 1871, that: "... in a few years, all great physical constants will have been approximately estimated, and ... the only occupation which will be left to men of science will be to carry these measurements to another place of decimals."

In fact, Maxwell strongly disagreed with such views and was attacking them. As it turned out later, scientists would not only fix the values of physical parameters, but would also discover the many concepts that ushered in twentieth-century physics.

Figure 2.1 James Clerk Maxwell (1831–1879). (Cavendish Laboratory, University of Cambridge.)

## A 'SELECT' LIST

The list we are looking for will therefore reflect present-day science. Remember, too, that its content will vary according to the field of specialty of the book consulted. A work on hydrodynamics will make reference to the density and viscosity of water, but a textbook of atomic physics will not. A textbook on relativistic physics will replace certain constants with equivalent ones: for example, the electron rest mass $m_e$ will give way to the rest energy of the electron $m_e c^2$. In this particular case, it is easy to make the transition from one constant to the other (simply by multiplying or dividing it by the speed of light squared), but it must be remembered that the list of constants is not absolute. It depends on the way in which we intend to use it.

Commenting on this fact in 1983, Steven Weinberg, an American physicist, proposed the improvement of a selective list, preserving only the most *fundamental* constants. By 'fundamental' he meant those that cannot be calculated on the basis of other constants, "... not just because the calculation is too complicated (as for the viscosity of water) but because we do not know of anything more fundamental."

Compiled in this way, the list of fundamental constants melts away like snow in the Sun. Immediately, we have to dispense with *g*, since, in Newtonian mechanics, acceleration due to gravity is expressed with reference to the gravitational constant *G*, to the mass and radius of the Earth, and to the latitude of the observer. Although *g* was a major player in Galileo's list of fundamental constants (if he had reasoned using modern terminology), it would be replaced by *G* in Newton's list, for his theory of gravity made *g* a derived parameter. Our list now consists only of those constants that cannot be explained in terms of more fundamental constants, which must be ascertained by measurement. These fundamental constants seem therefore to be "constants of Nature".

## THE LIMITS OF EXPLANATION

When he put forward this definition of a fundamental constant, Weinberg was expressing a change of viewpoint between the nineteenth and twentieth centuries. The list of constants reflects not only our knowledge, but more importantly our lack of knowledge. Each constant indicates the limits of explicability of a theory. Newton's theory of gravity links gravitational force to the masses of, and distance between, two interacting bodies, but it does not explain why the coefficient of proportionality, *G*, has its measured value. Neither does it propose some evolutionary law for *G*, which means that we must consider the coefficient as a constant. Indeed, Weinberg stated in 1983:

> "Each constant on the list is a challenge for future work, to try to explain its value."

No longer do we merely ascribe a value to constants: we also seek to explain them.

So, for each constant there are two situations: either it is fundamental, in which case we can only measure its value; or it is not fundamental, and we might harbour some hope of formulating a more general theory capable of predicting its value.

The acceleration due to gravity, *g*, has been removed from the list of fundamental constants as it belongs to the second category: Newton's gravitational theory, more general than Galileo's theory of falling bodies, allows us to calculate it. *G*, the gravitational constant, will remain a fundamental constant, until new knowledge comes along to deprive it of its status. This could happen, if a theory even deeper than general relativity were to replace *G* with another, even more fundamental constant. Or perhaps *G* might fall from the list if it were detected experimentally that there were some faint variation in its value, a variation hitherto hidden from view by the lack of precision in our measurements. Such a discovery would oblige theoreticians to find a new equation to describe the evolution of *G*. Note that the reverse could also be true: a variable might join the list of constants, as occurred with *c*, the speed of light.

So our rapid overview of the history of three constants, *G*, *g* and *c*, reveals the existence of a hierarchy of constants: they are not all as fundamental as we might think, nor do they all exhibit the same level of generality. So we are able to categorise them. It is likely that such a categorisation will itself evolve as our knowledge of physics increases.

## A CLASSIFICATION OF CONSTANTS

In 1979, physicist Jean-Marc Lévy-Leblond was the first to suggest a classification of fundamental constants, in three categories:

> In order of increasing generality, I distinguish:
>
> A. The properties of particular physical objects. Today, these are considered as fundamental constituents of matter; for example, the masses of "elementary particles", their magnetic moments, etc.
> B. The characteristics of classes of physical phenomena. Today, these are essentially the coupling constants of the various fundamental interactions (strong and weak nuclear, electromagnetic and gravitational), which to our present knowledge provide a neat partition of all physical phenomena into disjoint classes.
> C. Universal constants, i.e. constants entering universal physical laws, characterising the most general theoretical frameworks, applicable in principle to any physical phenomenon; Planck's constant, $h$, is a typical example.

To these we must now add a fourth:

> D. Reference constants, whose numerical value is fixed 'by decree', and which lie within the definition of our systems of units.

The speed of light ($c$) is a good example, entering this category in 1983. Since that date, this constant has been used in the definition of the metre. It is probable that, in the future, other constants will be used to define units.

We have already mentioned the link between the numerical value of constants and systems of units. In order to bring out the importance of this fourth status for fundamental constants, let us look briefly at the history of the metre.

## ONE WEIGHT, ONE MEASURE

Measurement consists of comparing the system studied with a system of reference, and it is the system of reference that defines the units of measurement. How is it chosen?

Like the history of physics, the history of units of measurement is marked by a quest for universality. Originally, human beings quite naturally used their own bodies to measure the lengths of the objects around them. The foot and the thumb gave us feet and inches, even though, of course, we do not all possess thumbs and feet of the same length. Normally, the bodily dimensions of some famous or outstanding personage, such as a king or a hero, would be the standard.

Albert Uderzo and René Goscinny introduced some humorous 'palaeometrology' into their *Astérix at the Olympic Games*, when they mentioned that the 600-foot running track at the stadium at Olympia was 197.7 metres long. Since the reference unit was the foot of Heracles, they wrote, "we can work out that the demigod took about a size 46".

Other old-fashioned units can seem surprising to us nowadays. For example, the 'span' was the measure of a hand with fingers fully splayed, from the tip of the thumb to the tip of the little finger. Similarly, the yard represented the distance between the end of the nose and the end of an arm and hand held out to the side. Volumes and material quantities were also measured in similar anthropocentric ways, and our receptacles became units in their own right: the basket, the barrel, the bushel, the pint ... Weights were related to what could be carried in the hand: the pound, the ounce, the dram ...

Obviously, this system had a few disadvantages. Each country or region, and each profession, had its own system of measurement. In France alone, there were more than 800 different units in use in 1789. When taxes were collected and business transacted, there was scope for much confusion, fraud and double-dealing. It was said that, in France, "each *seigneur* has a different bushel in his own jurisdiction" (*Cahier de doléances*, 1789). The worn-out kingdom could no longer tolerate this situation, and the call among the people was for "one king, one law, one weight, one measure".

From the seventeenth century onwards, the need was felt for a system of measurement based upon a natural standard. In 1670, Gabriel Mouton set out to define a unit of fixed length, the *mille*, as the length of one angular minute of a great circle on the Earth's surface. In 1671 and 1673, respectively, the Abbé Jean Picard and the Dutchman Christiaan Huygens chose as a reference the length of a pendulum that oscillated once a second. It had in fact been known since Galileo's time that the square of the period of oscillation of a pendulum is proportional to its length, and does not depend on its mass. However, the astronomer Jean Richer, who was conducting a geodesic survey at Cayenne in 1671, stated that a one-second pendulum was shorter in the Tropics than in Paris because $g$, the acceleration due to gravity, varies with latitude. By this statement the universality of the proposed unit seemed compromised. Prieur de Vernois, a military engineering officer, suggested that a fixed location for the pendulum should be specified, and he chose the Paris Royal Observatory. He also proposed that a platinum rod be fashioned to the same length as the pendulum, to serve as the standard length.

In spite of these trials and propositions, arbitrary and anthropocentric units held sway until the Revolution. The scientists of the nascent Republic really embraced the problem in a climate of universalist ideals such as equality: their cry was "one system for all, and for all time". This ideal system should be objective, not anthropocentric. Their aspirations were certainly reflected throughout society. The people yearned for a more equitable taxation system, and a unified set of weights and measures. Intellectuals and authorities dreamed of a universal system that would be accepted beyond the national frontiers, proclaiming the importance and influence of France. These demands came before the Assemblée Nationale with the help of the clergyman Charles-Maurice Talleyrand. On 8 May 1790, the Assemblée adopted the principle of a uniform system of weights and measures. It turned to the Académie des Sciences, where the best minds of the day (such as Jean-Charles Borda, Marie Jean Antoine de Condorcet, Louis de Lagrange, Pierre Simon de Laplace and Gaspard Monge) were asked to choose a unit of length based on one of three possibilities: (1) the length of the one-second pendulum; (2) a fraction of the equator; or (3) a fraction of a terrestrial meridian. It seemed that planet Earth would replace the human being, becoming the new reference for measurement.

## THE DECIMAL SYSTEM

The one-second pendulum was not acceptable to all, for the reason already evoked: it was not really universal. The equator also seemed too restrictive an idea, traversing as it does only some of the countries of the world, and seeming to depart from the notion of the equality of nations. The meridian was more promising – a universal line, since every country is crossed by a meridian, and all meridians are equal (Figure 2.2). Such was the consensus of the Académie on 26 March 1791, and Condorcet pointed out that "the Académie has sought to

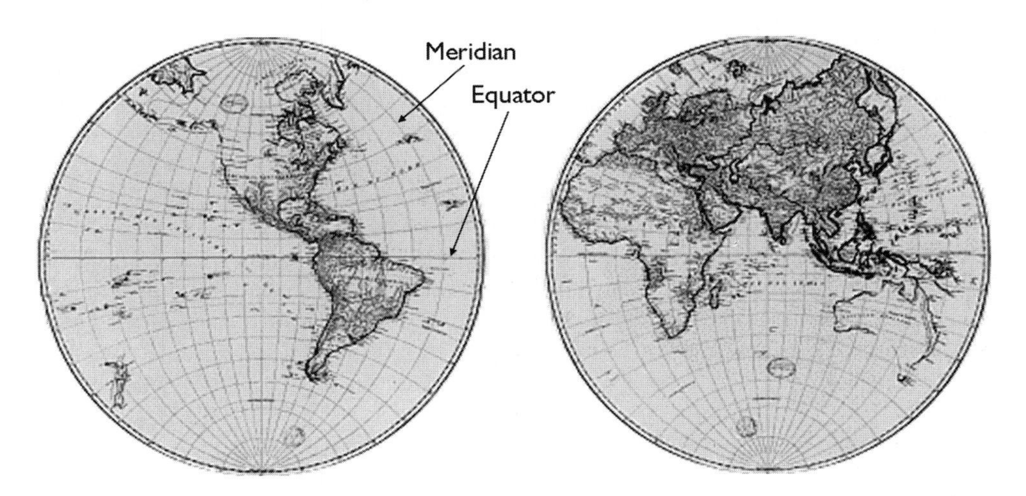


Figure 2.2 The Earth: its meridians and equator.

exclude any arbitrary condition that might cause some to suspect the influence of a certain self-interest for France".

On 30 March 1791, a decree was passed, stating that a quarter of the terrestrial meridian would be the basis for the definition of the 'metre' (from the Greek *metron*, a measure), as proposed by Borda. A metre would henceforth be one ten-millionth part of a quarter of the terrestrial meridian.

Jean-Baptiste Joseph Delambre and Pierre Méchain were given the task of measuring the length of the meridian between Dunkirk and Rodez, and Rodez and Barcelona respectively, and began this enterprise on 25 June 1792. Punctuated by various upheavals in an unstable period blighted by war, it took six years to complete. Provisionally, the Académie adopted as its unit of length the metre as defined from the measurement of the Paris meridian by Abbé Nicolas Louis de Lacaille in 1740. All these measurements used as a reference the 'Peruvian *toise*', defined on the basis of meridian measurements carried out in 1736 by Charles-Marie de la Condamine during his expedition to Peru.

The metre was then defined as 3 *pieds* (French feet) and 11.44 *lignes* of the Peruvian *toise* (or *toise de l'Academie*), at a temperature of 13 degrees Réaumur. (There were 144 *lignes* in 1 *pied*, so a *ligne* was about one-twelfth of an inch. The *toise* was 6 *pieds*, i.e. about 1.949 metres.) This value for the metre would be repealed by the Law of 19 Frimaire Year VIII (10 December 1799) which fixed a definite value of the metre of 3 *pieds* and 11.296 *lignes*. The accuracy of the measurements made by Delambre and Méchain was 15 times better than that of the Peruvian results.

To facilitate the measurement of objects to be studied, the metre is multiplied or divided by multiples of 10, providing a decimal system that responds to the necessity for simplicity. The universality of the denominations is respected by using the 'root languages' of Greek and Latin: the multiples of units (10, 100, 1000) have Greek prefixes (deca-, hecto-, kilo-), and the divisions (1/10, 1/100, 1/1000) are derived from Latin (deci-, centi-, milli-).

The unit of mass, the kilogramme, was linked to the metre by a definition given by the chemist Antoine-Laurent Lavoisier: it is the mass of a cubic decimetre of distilled water, at a given temperature and pressure. The mean solar day, which is the time that elapses between successive culminations of the Sun, lasts 24 hours. Each hour is divided into 60 minutes, and each minute into 60 seconds. There were certain attempts to apply the decimal ideal to the measurement of time, but any attempts at decimalisation were fiercely resisted. During the Revolution, scientists divided the day into 10 'new hours', the hour into 100 'new minutes', and the minute into 100 'new seconds', but there was such an outcry that the idea was abandoned in 1795. The familiar old clock-faces, with their 12 hours and 60 minutes, continued to serve.

## AN INTERNATIONAL SYSTEM

Now that the new units had been defined and the meridian measurements settled, Talleyrand invited France's allies and neutral nations to take part in the

Figure 2.3 (a) Modern standard metre and kilogramme (BIPM). (b) Standard metre in the Rue Vaugirard, Paris.

determination of definitive standards (Figure 2.3). Two platinum standards were manufactured: a standard metre bar, and a standard kilogramme cylinder. These 'prototypes' were lodged in 1799 in the Republican Archives, in a metal cabinet, at the Pavillon de Sèvres near Paris (Figure 2.4).

Paradoxically, the metric system was adopted more readily in some other countries than in France – the United States being one of the first nations to accept it. The German physicist Carl Friedrich Gauss was using the system in 1832, while in France the new units were still, sadly, not yet in common usage.

Even more sadly, the Napoleonic Empire began to neglect the ideals of the

Figure 2.4 The Pavillon de Sèvres near Paris, home of the international standard metre and kilogramme. (© BIPM.)

Revolution and authorised the use of old traditional units in 1811. The *toise*, the *aune*, the bushel and the ounce came back into their own, sowing confusion in their wake. At last, on 4 July 1837, a law was passed imposing the metric system to the exclusion of all others. Decades later, the decimal metric system was accepted, and by the mid-nineteenth century it was in use in many countries. Commercial exchanges could be then transacted on a basis of common values. In 1875, 17 countries signed the Metre Convention (*Convention du Mètre*), leading to the creation of the Bureau International des Poids et Mesures (International Bureau of Weights and Measures: BIPM), the international metrology centre at Sèvres, to the south-west of Paris. Here, world standards for physical measurement are monitored, and prototypes kept. In 1889, the BIPM made a new platinum–iridium prototype, containing a small amount of the very dense metal iridium. The metre would henceforth be defined as the distance between two marks etched on the prototype, to within a few tenths of a millimetre. The standard now became accurate to one part in 10 000 ($10^{-4}$).

## THE ATOM: A UNIVERSAL REFERENCE

Even though great progress had been made towards universal standards and greater accuracy in measurement, there were still some problems to be overcome with the metric system as the nineteenth century drew to a close. The prototypes of the metre and the kilogramme were stored in a single location, at Sèvres, and scientists from other countries had to make long journeys to calibrate their secondary standard bars or make copies. Moreover, the reference objects were not immutable: wear and temperature variations could affect them in different ways at the different locations in which they were kept. The definitions of the units themselves were still by no means universal, as James Clerk Maxwell pointed out. In 1870, he remarked during a presidential address to the British Association for the Advancement of Science that the Earth remained an arbitrary and inconstant reference. As the Earth evolved, it might contract through cooling, or expand as meteoritic material accumulated upon it: the metre would therefore be modified. Also, the Earth's rotation might slow or accelerate, modifying our definition of the second (current estimates tell us that the length of the terrestrial day is increasing by two thousandths of a second per century, because of tidal effects). The terrestrial environment, therefore, was not stable enough to serve as a reference framework for standardisation. Furthermore, these standards were still anthropocentric: they could not be communicated to some intelligence on a different planet. In order to make these standards independent of time and our position in the universe, we should base them on the most universal laws of Nature.

Maxwell was then studying the kinetics of gases, describing a gas as a collection of microscopic elements: atoms or molecules. He realised that atoms of the same element might constitute universal references, since they all have exactly the same mass and vibrational periods. To understand how Maxwell's idea might lead to new standards of measurement, we must examine what he meant by "the period of vibration of an atom". Therefore, a little detour into spectroscopy is necessary ...

In the early nineteenth century, the German physicist Gustav Kirchhoff invented the spectroscope. He enclosed gas in a tube and subjected it to electrical discharges. The gas emitted light, which he directed to pass through a prism. In his day, Newton had also used prisms, to split white light into a range of rainbow colours. Kirchhoff discovered that pure chemical elements, such as hydrogen, nitrogen, oxygen and sodium, did not emit a spectrum of colours, but bright lines of distinct colours, separated by dark zones. He found, for example, that sodium emits an orange-yellow line in visible light, known as the 'D line', and that sodium is the only element to emit this particular line. (The "D-line" is actually a doublet of lines with wavelengths of 588.995 and 598.592 nm.) Each element can therefore be characterised by its specific emission lines, just as fingerprints can identify individual people. Being unaware of the processes that caused atoms to emit or absorb light, scientists presumed that the bright lines resulted from vibrations peculiar to each atom, giving rise to Maxwell's idea of "the period of vibration of an atom".

Maxwell therefore imagined that it might be possible to use the properties of atoms to define units that were truly universal, and his ambition was realised nearly a century later when developments in atomic physics and spectroscopy achieved the universal, objective standard that he had envisaged. In 1960, the BIPM established a new definition of the metre based on the emission of light by atoms, taking care to ensure continuity with the existing metre. The metre is defined as the length equal to 1 650 763 wavelengths, in a vacuum, of radiation based upon the transition between the levels $2p_{10}$ and $5d_5$ of the atom of krypton-86.

Thus defined, the new metre is compatible with the standard metre in its interval of uncertainty, but its accuracy is increased by a factor of 10 000, or one part in 100 million ($10^{-8}$).

## MEASURING TIME

The unit of time evolved together with the metre. The disadvantage of defining the second astronomically, by linking it with the motion of our planet, is that Earth's rotationary period is not fixed. For example, tidal forces are slowing the rotation of the Earth on its axis, causing the length of the day to change as the centuries pass. In 1927, André Danjon, of the Strasbourg Observatory, proposed that the rotation of Earth as a reference should be abandoned and replaced by the revolution of the planets around the Sun, thereby using the year rather than the day as a basis for the measurement of time. In 1952, the BIPM defined the second as "1/31 556 925.9747 of the tropical year", a tropical year being the time that elapses between one vernal equinox and the next. The accuracy of this new astronomical second was a few parts in 100 million. However, it still depended on astronomical observations that were not only time-consuming and laborious, but remained subject to the various irregularities of celestial mechanics.

Maxwell noted that the vibrational period of atoms could also be used to measure time, and this idea has led to the elaboration by scientists of atomic clocks, using their favoured element (Chapter 5 explains why), the caesium atom, which vibrates at almost 10 billion oscillations per second. The accuracy of the second was enormously increased. Physicists also noted that atoms oscillate with a regularity that no mechanical clock achieves, and they seriously considered replacing the 'astronomical second' with the 'physical second'. In 1967, the BIPM therefore decided that the second should be "the duration of 9 192 631 770 periods of the radiation corresponding to the transition between the two hyperfine levels of the ground state of the caesium-133 atom".

Here again, the new definition respects the old, but the accuracy is now within a few parts in 10 billion ($10^{-10}$). Progress in the science of atomic clocks has meant that the second is now the most precisely known of all units, with a relative accuracy of 1 part in 10 000 billion (or 10 trillion, $10^{-13}$).

This great precision led the BIPM to modify, yet again, the definition of a metre, which now seemed a rather poor relation of the second. In 1983 the Bureau fixed the value of the speed of light, and the metre became a unit derived from the second. The new definition read:

> "The metre is the length of the path travelled in a vacuum by light during a time interval of 1/299 792 458 seconds."

The metre had now achieved an accuracy similar to that of the second, and both units were linked to the speed of light – a fundamental constant. We now see a new role for fundamental constants in physics: that of defining units. Given the importance of measurement and units in physics, it became necessary to create a category (designated D) for constants used in this way, even if, for the time being, the speed of light is the only member of this very exclusive club.

## THE EVOLUTION OF CONSTANTS

We are beginning to see the world of constants a little more clearly. The list and the classification of fundamental constants reflect the state of knowledge at a given epoch. Each new step forward in our comprehension of physical phenomena introduces an evolution, or change of status, of one or more of these constants. Although the changes of status may be of different kinds, let us examine these evolutionary processes with the aid of a few examples.

The *emergence* of a constant often accompanies the discovery of a new system or new laws: $g$ and $G$, for instance, stem from the laws of Galileo and Newton. This was also the case with Planck's constant in 1900.

An existing constant may also move from one class to a more universal class: such a change of status is known as a *progression*. An example is that of the electric charge, which was originally considered to be a characteristic of only the electron (category A), but was regarded as more general when it was discovered that the proton, although much more massive than the electron, bore exactly the same charge, with opposite sign. The discovery of other charged particles, e.g. the muon, confirmed the 'granular' character of electricity, the charges of all free particles being simply multiples of the electron charge (or 'elementary charge'). Consequently, this charge characterises the electric interaction between all charged particles (category B).

Conversely, a constant may reveal itself to be less universal than was once thought, and suffer a *retrogression*. The evolution of $G$ might well be described as such, even if this (*a posteriori*) description seems a little artificial. In Newton's time, physics confined itself to the description of motion and gravity, which conferred a considerable generality upon these phenomena. If our classification had been carried out then (and if the gravitational constant had been defined), $G$ would have been in category C. However, after optics, magnetism, electricity and thermodynamics had been studied, gravitational effects became just another particular class of phenomena. We acknowledge this fact by 'retrogressing' $G$ from category C to category B.

Another change of status is by *transmutation*, where one constant is replaced by a more general one. For example, the electric charge, representing the strength of electrical interaction (category B) gives way to the coupling constant of the so-called electroweak interactions, following the unification of the electromagnetic and weak nuclear interactions.

Finally, a constant may quite simply disappear from the list of fundamental constants for a variety of reasons, which are worth exploring.

## THE DISAPPEARANCE OF CONSTANTS

We can distinguish four causes for the *disappearance* of constants from the list: fusion, explanation, becoming a variable, or becoming a conversion factor.

We encountered a case of fusion in Chapter 1, with Galileo's law of falling bodies and Kepler's third law on the elliptical orbits of the planets. With our mathematical approach, we see that these laws involve two constants, the acceleration due to gravity $g$ and the constant of proportionality $K$ between the square of the orbital period and the cube of the semi-major axis of the ellipse. Newton's elaboration of the theory of gravity led to the disappearance of these two constants in favour of a unique concept represented by the gravitational constant $G$.

Subordination (explanation) is an obvious cause of the disappearance of a fundamental constant. As soon as a constant can be explained as a function of more fundamental constants, it is banished from the list. A good illustration of the subordination of a constant is furnished by the study of the black body, which we shall examine more closely later when discussing Planck's constant. Before Planck, in 1893, German physicist Wilhelm Wien had made an important advance with his proposition of an empirical function to describe the spectrum of the black body. This function involved two constants, called $a$ and $b$. Determined by measurement, these constants were considered to be fundamental, since no theory could yet explain them. Revising the problem from first principles, Planck was to provide a complete solution, and introduce a new constant, $h$. From that moment, constants $a$ and $b$ lost their *raison d'être*.

A constant may also disappear because it is found to be a variable. This has not yet occurred, but following the example of John Webb's team, physicists have begun to scrutinise constants with a view to verifying their stability through time with ever greater accuracy. If, one day, one of them shows some variation, however infinitesimal, it will be necessary to formulate an equation that explains its evolution. It will then cease to be a constant, and will become a dynamical variable.

Finally, it can happen that a better understanding of physics teaches us that two hitherto distinct quantities can be considered as a single phenomenon. Such was the case in 1842, for example, with heat and work. In that era of great industrial progress, it was important to know how to transform the heat in a boiler into useful mechanical work in machines. In Britain, James Joule realised that these two quantities were in fact one: they were two different forms of energy. The equivalence of work and heat is the basis of the first principle of thermodynamics: work can be transformed into heat and *vice versa*.

Joule's constant, expressing the proportionality between work and heat, lost any physical meaning and became a simple conversion factor between units used in the measurement of heat (calories) and units used to quantify work (joules). Nowadays, hardly anyone uses this conversion factor, as work and heat are both expressed in joules, and the calorie has fallen into disuse.

## THE INQUIRY GOES ON

We now have the necessary knowledge to pursue our inquiry into fundamental constants. We know how to list them, and classify them in order of importance. On the basis of a few examples, we have defined categories of fundamental constants and understand the processes by which they may move from one category to another. Most importantly, we are now in a position to be able to recognise the most fundamental constants. The most universal categories, which we have called C and D, currently contain only three constants: the speed of light $c$, the gravitational constant $G$, and Planck's constant $h$. It is safe to say that their role in physics, judging by their exceptional status, is of prime importance. Their role will be the next subject for our commission of inquiry.

# 3

# Planning the edifice: structure of theories

Our commission now turns its attention to the three pillars that seem to give physics its structure: the speed of light *c*, the gravitational constant *G*, and Planck's constant *h*. These are the fundamental constants in the most universal of our categories. In order to understand their exceptional roles, we need to spend a little time on their evolution within the history of discovery and progress in physics. This will lead us to an inventory of the edifice of physics, which we will use to reveal its weak spots, where cracks might appear if constants are not as constant as they seem. Let us now examine our knowledge of the evolution of physics!

## FAST, BUT THERE IS A LIMIT

We begin with the speed of light. It was long believed that light propagated *instantaneously* through air and in a vacuum. Galileo, who would not accept such dogmatic beliefs *a priori*, set out to test this idea. He tried to time the return journey of light between two hills a few kilometres apart. At night, he showed a lantern, having instructed an assistant to reveal his own lantern only when he saw the light of the first. Across that distance of a few kilometres, the light did in fact seem to travel instantaneously; not surprising, since the travel time for the light between the two hills was of the order of one hundred thousandth (0.00001) of a second – an interval that could not possibly be measured at the time.

Later, French philosopher René Descartes saw that the distance involved in such an experiment might be increased by using the Earth and the Moon as reference points. If there was a time delay between the moment when the Sun, the Earth and the Moon aligned and the moment when a total eclipse was actually observed, it might be possible to calculate the speed of light. Descartes found no such delay, and firmly concluded that light propagated instantaneously..

It was indeed fortunate that later researchers were not discouraged by the findings of Galileo and Descartes. The Danish astronomer Øle Roemer (Figure 3.1), who had been invited to the Paris Observatory in 1671, further increased the reference distance for measuring the speed of light by considering the

Figure 3.1 Øle Roemer and his telescope.

eclipses of the satellites of Jupiter. He found that the eclipses of the satellite Io were occurring slightly later or sooner than had been predicted by ephemerides provided by his patron, Jean-Dominique Cassini. 'Late' eclipses happened when the distance between the Earth and Jupiter was increasing, and 'early' eclipses occurred when the distance was decreasing. Roemer reasoned that this must be because light takes a certain time to travel here from the vicinity of Jupiter, a fact that had not been taken into account in the computation of the ephemerides.

Our knowledge that light travels at a finite speed can therefore be said to date from 1676, when Roemer published his results. Henceforth, the existence of the speed of light in a vacuum was just as much a quantity as the speed of sound in the air. These speeds were both particular phenomena in category A of our classification of physical constants.

English astronomer James Bradley provided a further estimate of the speed of light in 1728 during a study and interpretation of what is now known as the aberration of starlight when he observed the star Gamma Draconis. This star showed a displacement in the sky owing to the Earth's orbital motion. The study of the apparent displacement of the star was of three-fold significance: it supported Copernicus' theory that the Earth moved around the Sun; it confirmed the finite speed of light as demonstrated by Roemer; and it provided a method for estimating the speed of light, which was later found to be about 10 000 times the orbital speed of the Earth. Unfortunately, this estimate also employed the distance from the Earth to the Sun, which at that time was not well known.

A century later, using ingenious instruments with rotating mirrors, Armand-Hippolyte-Louis Fizeau and Léon Foucault, in 1849 and 1862 respectively, determined the speed of light in a vacuum in Earthbound experiments of great accuracy. They concluded that its value was approximately 300 000 kilometres per second. Although the speed of light was by then well known, it still resided in category A, since it was a characteristic of only the speed of light. It was still far from being a constant, since, according to Newton's optical theory, it could not be separated from the ballistic speeds of other bodies. Its value would depend on the speed of the body emitting the light, and on the motion of the observer. This question remained at the centre of scientific debate for decades afterwards.

## A WAVE...

The speed of light moved into a more universal category in the second half of the nineteenth century when the nature of light itself was more comprehensively understood. Light had long been a subject of debate; Isaac Newton had studied light by passing it through substances of various compositions and shapes and had used prisms to recreate the phenomenon of the rainbow indoors. Newton imagined light to be a corpuscular phenomenon, which travelled in a straight line, rebounding from reflecting surfaces and altering its trajectory as it passed from the air into another transparent medium. His Dutch contemporary, Christiaan Huygens, thought otherwise. He believed that light was a propagating wave, similar to the action of a pebble falling into water, creating surface ripples that spread in all directions from the point of impact. Huygens perceived that, from its source, light propagated in all directions through space – an idea that would explain the phenomenon of the diffraction of light through a diaphragm. Diffraction occurs when a wave encounters an obstacle of approximately the same size as its wavelength (the distance between two successive waves); in this case, the wave seems to be retransmitted in all directions (Figure 3.2).

These two rival theories of the nature of light, as corpuscular or as a wave, coexisted for two centuries. In the early nineteenth century, the supporters of the wave theory seemed to have supremacy following experiments by Thomas Young in England and Augustin Fresnel in France, who showed that light exhibits

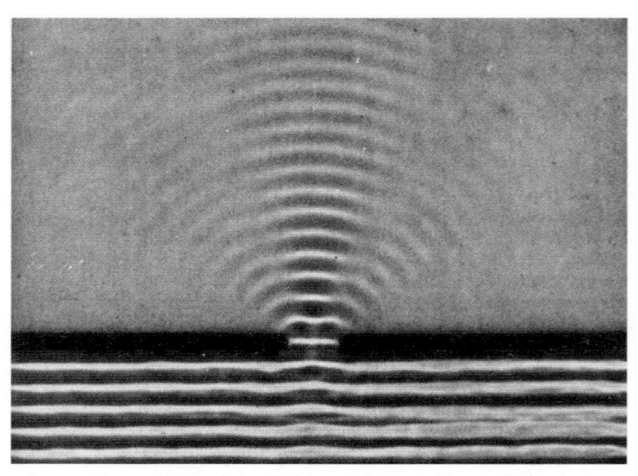

Figure 3.2 Diffraction of a wave on the surface of water, caused by a slit in a barrier. (After *PSSC Physics*, Haber-Schaim, Doge, Gardner and Shore, Kendall/Hunt Publishing Companies, 1991.)

interference effects that they attributed to its wave-like character. If we throw two pebbles into a pond, circular ripples will spread from their points of impact and will meet. When the wave-crests meet, they will combine to form an even higher crest, but when this crest meets a trough, they cancel each other out. The ripples then create a different succession of troughs and crests: an interference pattern (Figure 3.3). In the case of light, the interference pattern shows alternating dark and light fringes (Figure 3.4). The fact that light is diffracted, and shows interference, proves that it travels in waves, just like those on the surface of water. What, however, is the nature of these luminous waves?

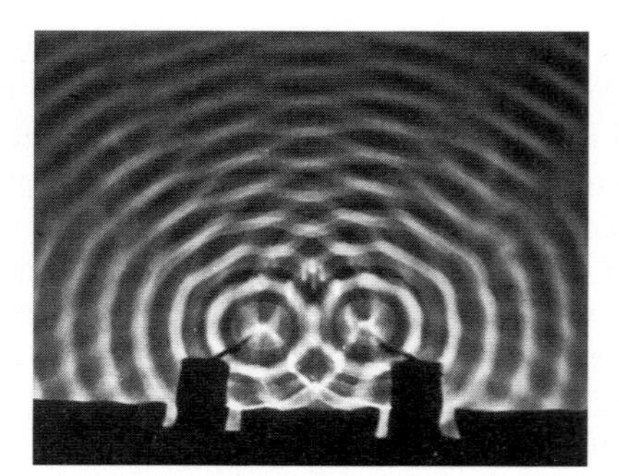

Figure 3.3 Interference pattern on water.

Figure 3.4 Interference pattern of light.

## ...AND AN ELECTROMAGNETIC ONE

James Clerk Maxwell determined the nature of light with a brilliant elucidation, which is still admired by many physicists. Magnetism and electricity had long been separate fields of study, but, in the early nineteenth century, the link between them was becoming apparent. In England, Michael Faraday showed that a wire carrying an electric current caused the needle of a compass to deviate. Conversely, a variation in a magnetic field induced an electric current in a nearby conductor. It therefore seemed possible that electricity and magnetism could be unified within a single description, which is what Maxwell accomplished in 1873. He reviewed various laws from the past, reformulated them and presented them as a coherent whole with four equations on which the theory of electromagnetism now rests.

Maxwell transformed a master-stroke into a stroke of genius by predicting, on the basis of his equations, waves of a new kind: *electromagnetic waves*. From the point of emission, variations in electrical and magnetic fields will propagate in all directions in space. The equations showed that the speed of this propagation in a vacuum was similar to that of light, as determined by Fizeau and Foucault. Kirchhoff, noticing this coincidence of values, saw that it might reflect the existence of a law of physics. Maxwell then inferred that light waves were electromagnetic, thus combining three concepts into one! Nearly 60 years later, Dirac would try (as already mentioned) to draw conclusions from another numerical coincidence.

The reality of electromagnetic waves would be confirmed experimentally in 1887, eight years after Maxwell's death, when Heinrich Hertz in Germany succeeded in creating them with an electrical oscillator. The characteristics of these waves (their propagation, reflection, refraction and interferences) resembled those of light, which agreed with Maxwell's theories. Only its wavelength – just thousandths of a millimetre – distinguished light from other Hertzian waves, which could be longer than a metre, or even a kilometre. The concept of the electromagnetic wave now brought together a whole class of phenomena, characterised by a single speed of propagation in a vacuum, $c$; it is a worthy member of category B in our classified list of constants..

## THE INVISIBLE ETHER

The constant $c$ was further promoted only 30 years after Maxwell. This resulted from a conceptual difficulty concerning the propagation of light. Physicists had imagined that some medium was needed to support this propagation, just as the surface of water was required to support waves. As the supposed medium, known as the 'ether', was thought to fill the universe and carry the light of the stars to our eyes, efforts were made to describe its properties. It was considered that since the Earth moved in an orbit, there ought to be an 'ether wind' effect, which might be detectable, and to this end the American scientist, Albert Michelson,

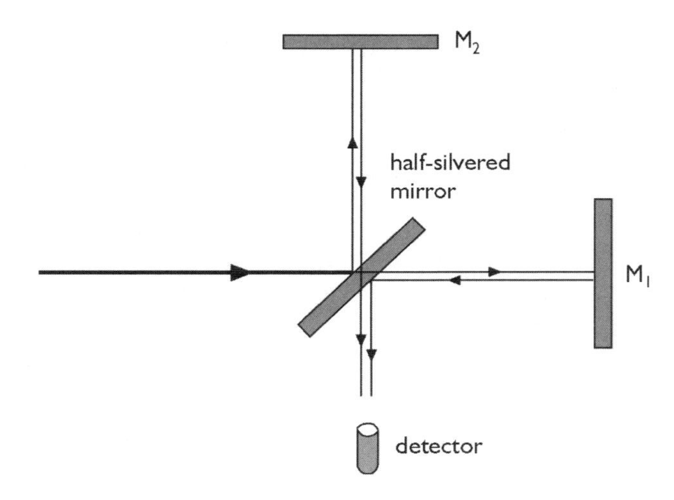


Figure 3.5 Michelson interferometer. A beam of light is split into two by a semi-transparent mirror (centre). One beam propagates in the direction of mirror $M_1$, and is then reflected from it back towards the partial reflector, which turns it towards the observer. The other beam is first sent in a direction at right angles to the first, and is then reflected by mirror $M_2$ towards the observer, who examines the interference between the two beams. If an interference pattern appears, then the lengths of the two paths are different.

proposed to investigate the ether, using the phenomenon of the interference of light waves. If a beam of light is separated into two, by passing it, for example, through two parallel slits, we obtain two identical sources which cause interference on a screen. This was how Young had proved that light creates interference patterns. Now, if one of the beams follows a longer path than the other, the interference pattern will be modified, and the dark and bright fringes will be shifted with reference to those of the original pattern.

Michelson's idea was to send a beam of light in the direction of the 'ether wind', while the other beam would travel in a direction perpendicular to it (Figure 3.5). At the end of their trajectories, each of the two beams would encounter a mirror, and be reflected back towards their source, where the experimenter would cause them to produce interference. If one of the beams, drawn along by the ether wind, were travelling faster than the other, the interference fringes would appear shifted with reference to the original pattern. Working sometimes with his collaborator Edward Morley, Michelson performed this experiment in interferometry several times until 1887. Unfortunately, his results were negative: light travelled at the same speed in all directions. The 'ether wind' did not exist!

Nor did the ether exist! Or so, in 1905, said Albert Einstein, who was an employee of the Bern patent office and unknown within the physics community. Nothing prevents light, or rather the oscillations of magnetic and electrical fields, from propagating without support. Also, the Michelson–Morley experiments showed that the speed of light does not depend on the frame of reference in which it is measured. At last, $c$ became a true constant, proper to both light and electromagnetism, and valid for all inertial frames.

It seemed, however, that the constant nature of the speed of light contradicted one of the rules of Galilean mechanics. According to Galileo, when two observers in relative, rectilinear and uniform motion observe the same phenomenon, the velocities measured by one of them are equal to the velocities measured by the other, to which is added to the relative speed of the two frames of reference. This simple transformation allowing us to pass from one inertial frame to another does not apply with light. We cannot add the speed of light to the reference frame of another observer.

Instead of Galileo's transformations, Einstein had to adopt new 'Lorentzian' ones, based on the earlier work of the Dutch physicist Hendrik Antoon Lorentz. These transformations involve, simultaneously, the speed of the moving reference frame and the speed of light. They take into account, mathematically, an effect now confirmed daily in particle accelerators. The inertial mass of a particle being accelerated increases as its speed approaches the speed of light. So, the applied force required to accelerate it also increases, and as the speed of light is approached, the force required to reach that speed becomes infinite. For this reason the speed of light is an inaccessible speed limit for a solid particle: it cannot be exceeded by the addition of another velocity.

On the positive side, Lorentz's transformations correspond to Galileo's as long as the speed of the system studied remains much less than the speed of light. Within this framework, the classical mechanics of Galileo and Newton remain valid, and have proved their worth in many domains. Einstein's new theory encompassed classical mechanics and also took account of the greater speeds of new, *relativistic* phenomena.

## NEW SYNTHESES

Einstein sacrificed Galileo's transformations in order to extend the principle of Galilean relativity: the laws of physics are the same in all inertial frames of reference. One of the foundations seemed fragile to him, while the other held firm. The latter would support a new relativistic mechanics (special relativity). The fact that light propagates at the same speed for two observers moving relative to each other has an important consequence: two events that appear to be simultaneous to one observer no longer appear to be simultaneous to the other. In other words, the measurement of time will not be exactly the same for both observers. Simultaneity was to Einstein what immobility was to Galileo: relative to the observer.

Herein lies the originality of Einstein's relativity: it makes time and space inseparable. Galileo's transformations dealt only with spatial coordinates, for example longitude, latitude and altitude on Earth. Lorentz's transformations brought in temporal coordinates, whose time was that as measured by a clock. In order to exchange information with a colleague who is a physicist travelling in a train at a constant velocity, an observer must specify both his own spatial co-ordinates and the time on the watch he is wearing. Physics now has its basis not only in spatial geometry, but in the geometry of a unified space and time – i.e. space-time. The speed of light characterises this space-time, so much so that it is involved in all the physical phenomena that unfold within it. The speed of light is also linked with notions of causality, in whatever physical system, which puts it into category C of our classification of fundamental constants. 'The speed of light' is no longer the best name for this constant, and the 'space-time structure constant' might be more suitable, given its universal nature. But this has even further significance ...

Constant $c$ synthesises the concepts of space and time. A few weeks after the publication of his article on the ether and the speed of light, Einstein realised that $c$ synthesises two other concepts: mass and energy. He established the formula $E = mc^2$, relating the rest energy of a body to its mass. This formula was to become the most famous in the history of physics, and we shall explore its consequences later, when we discuss nuclear physics.

In the previous chapter, we recounted the last act in the story of the evolution of the speed of light, when dealing with the subject of units of measurement. Atomic clocks had now achieved such accuracy in the measurement of time that the BIPM intended to apply this fact to the measurement of the metre. After all, we can use a time-period to describe a distance; ramblers discuss their journeys in terms of hours spent walking, and astronomers state the distances to the stars in light-years, one light-year being the distance light travels in a year. In 1983, the BIPM decreed that the value of $c$ should be 299 792 458 metres per second, allowing us to derive the value of the metre from the value of the second. The speed of light thus entered category D of our classification, and, for the present at least, it is the only constant there...

## *G* FOR 'GREAT LEAPS FORWARD'

The second fundamental constant whose evolution we shall discuss is Newton's gravitational constant $G$. We have already seen how $G$ was 'downgraded' from category C (the most universal constants) to category B (constants describing a class of phenomena). This progressive change of status was the result of the numerous subjects studied in physics in the eighteenth century. It was discovered that electricity and magnetism obeyed laws that involved constants other than $G$. These constants, describing classes of phenomena, also belong to our category B.

In the two centuries following Newton's work, there were many highly successful studies involving Newtonian gravity. Some researchers measured the value of the gravitational constant with ever-increasing precision, while others developed a mathematical formalism of the theory. Observers of the skies confirmed that the planets moved in agreement with the calculated orbits.

British scientist Henry Cavendish made the first accurate measurement of $G$ in

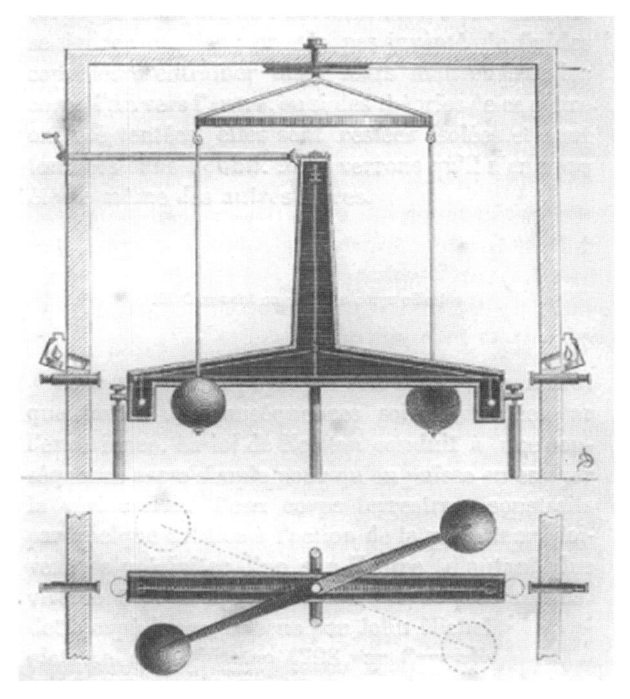

Figure 3.6 Cavendish torsion balance. Two masses are suspended from a rod which is free to turn about its centre (top). The angle of rotation of this rod (below, as seen from above) enables the force of the interaction between the masses and two other test masses that are moved closed to them, to be measured. (From *Les Forces Physiques* by Achille Cazin, Librairie Hachette, Paris, 1871 edition.)

1798. This fine experimenter used a very accurate apparatus: a torsion balance (Figure 3.6), consisting of a horizontal rod, with a metal sphere at each end, suspended by a wire. When two equal and opposite forces were applied to the spheres, the wire twisted by an amount that enabled the applied forces to be measured. Cavendish caused the wire to twist by placing two heavy lead weights near the balance. It was a simple enough experiment, but it must be remembered that the force exerted by a ball with a mass of 100 kilograms is about 100 000 smaller than the actual weight of the ball! In spite of this, Cavendish achieved some very accurate results. Transposed to a modern context, using the system of units (SI), his value for $G$ becomes $6.71 \times 10^{-11}$, which is quite close to its modern value of $6.67 \times 10^{-11}$: and this constant remains, even today, the most difficult to measure.

Cavendish was actually attempting to measure the density of the Earth rather than find $G$, which at that time was not of immediate interest to scientists. He was, in effect, 'weighing' the Earth. His result enabled astronomers to deduce the mass of the Sun and all the other planets under its influence through the application of Newtonian mechanics via Kepler's third law.

Astronomy made great progress thereafter. In 1781, William Herschel discovered Uranus, the third of the giant planets after Jupiter and Saturn, and in 1787 he found two of its satellites. Although the orbit of Uranus differed from that predicted by Newton's theory, it was nevertheless Newtonian mechanics that solved this puzzle when, on 31 August 1846, the French mathematician Urbain Le Verrier published the orbital characteristics of a hypothetical planet that would produce the observed effect on the orbit of Uranus. In England, John Couch Adams had performed similar calculations. On 23 September 1846, German astronomer Johann Galle discovered the planet Neptune by directing his telescope towards its calculated position, confirming in no uncertain terms the predictive power of Newtonian mechanics.

## GRAVITY AND SPACE-TIME

The real change in the status of $G$ had to wait until 1915, when Einstein presented his theory of general relativity, 10 years after the appearance of what is known as his special theory of relativity. The latter, as we have seen, relied on Lorentzian transformations, just as Newton's tried and tested theory of gravity was based upon Galileo's transformations. What new implications for the description of gravity arose from the new transformations?

Even if we feel the need to reform our theories, we ought not to throw out the baby with the bathwater, and regress to zero. We should instead build our edifice on solid foundations, even if they are old ones. Einstein not only had retained Galilean relativity in his elaboration of the special theory, but he relied on another Galilean principle, the universality of free fall, to underpin his reform of gravitational theory. This principle states that all bodies fall with the same acceleration in a given gravitational field (Figure 3.7). The brilliant mind of Einstein realised, however, that observers in free fall cannot perceive the gravitational field in which they are falling. This is of course a thought experiment: the illusion would be only a temporary one! In a lift in free fall, all the occupants would float in a state of weightlessness. Can we go as far as to claim that, in this frame of reference, the laws of physics do not involve gravity? Einstein would answer yes: locally, gravity can be cancelled out.

Pursuing his argument, Einstein proposed that the trajectories of bodies in free fall could be considered as paths of minimum length (geodesics) in a curved space-time. To help us to understand what this means, let us reduce the four dimensions of space-time to the two dimensions of a roller-skating rink. The ups and downs of the surface divert the trajectories of the skaters; in space-time, it is the presence of matter that causes curvature – the greater the mass, the greater the curvature. The local curvature of space-time determines the trajectories of bodies in the vicinity.

Having securely married space to time, Einstein then proceeded to link matter to them. The container and the contents then became one.

The linking parameter was the gravitational constant $G$, which simply

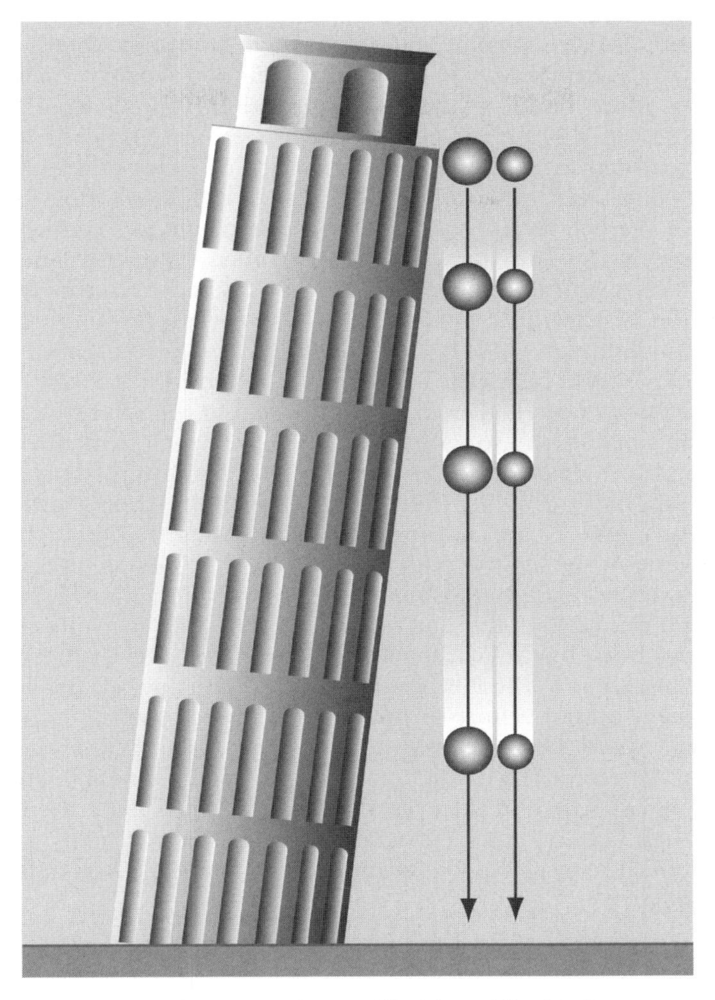

Figure 3.7 The Tower of Pisa experiment. All objects fall with the same acceleration, such that they hit the ground at the same time if dropped from the same height at the same speed.

expresses the capacity of space-time to bend in the presence of matter, whatever that matter might be. $G$ thus regained its universal significance, earning its place in category C.

## GENERAL RELATIVITY: FROM SUCCESS TO SUCCESS

Since its formulation, the theory of general relativity has been confirmed on many occasions, two of which will be described here. The first concerns the anomalous orbit of the planet Mercury, which Newton's theory could not explain. Mercury's perihelion, i.e. the point in its orbit where it is closest to the Sun, advances more rapidly than Newtonian theory predicts. The observed anomaly (43 seconds of arc per century) may seem insignificant to us, but it cannot be explained as a result of the presence of some other object in the vicinity. The discovery of another Neptune is excluded! Einstein's theory, however, does predict such a value: 43 seconds of arc per century. Mercury, which is closer to the Sun than the other planets, experiences more marked relativistic effects, and the 43 seconds of arc anomaly is the result of those effects. The application of Newton's theory is therefore seen to be limited to regions where the gravitational field is not too intense. On the other hand, the application of general relativity can now be extended and can be used to calculate corrections which, although infinitesimal in the solar system, could be of great significance in the study of massive objects such as neutron stars and black holes across the vast distances of interstellar and intergalactic space.

A second success of general relativity concerns light. If, as Pierre de Fermat claimed as long ago as the seventeenth century, light follows the shortest distance between two points, its path should be seen to deviate in the vicinity of a mass – for example, starlight passing close to the Sun should be displaced by 1.7 seconds of arc because of the Sun's mass (Figure 3.8). The physicist Arthur Eddington, impressed by the mathematical beauty of Einstein's theory, set off in 1919 on an expedition to the Portuguese island of Principe, off the west coast of Africa, to observe a total eclipse of the Sun. With the blinding rays of the Sun eclipsed, he measured the angle by which the light of stars in the same direction as the Sun was deflected. The prediction was confirmed. General relativity and its originator were vindicated in the eyes of the world. Since 1919, there have been many other such occasions, and general relativity now has many applications, ranging from our use of global positioning systems to our understanding of the workings of pulsars!

## GETTING TO GRIPS WITH THE BLACK BODY

The theory of relativity was not the only scientific innovation in the productive early years of the twentieth century. This era also saw the first stirrings of quantum mechanics and the introduction of Plank's constant, $h$. Unexpectedly, they sprang from the body of the 'goddess' thermodynamics, when she encountered the thorny problem of black-body radiation.

Every heated body, from the blacksmith's irons to the stars in the night sky, emits radiation. For example, the Sun, the source of our daylight and warmth, emits light in all the colours of the rainbow, as well as infrared and ultraviolet radiation. When these radiations are arranged according to their wavelengths (or frequencies), they form a spectrum (Figure 3.9). Can we use a theory to predict the spectrum of any heated body? Gustav Kirchhoff asked fellow theoreticians this question in the late nineteenth century.

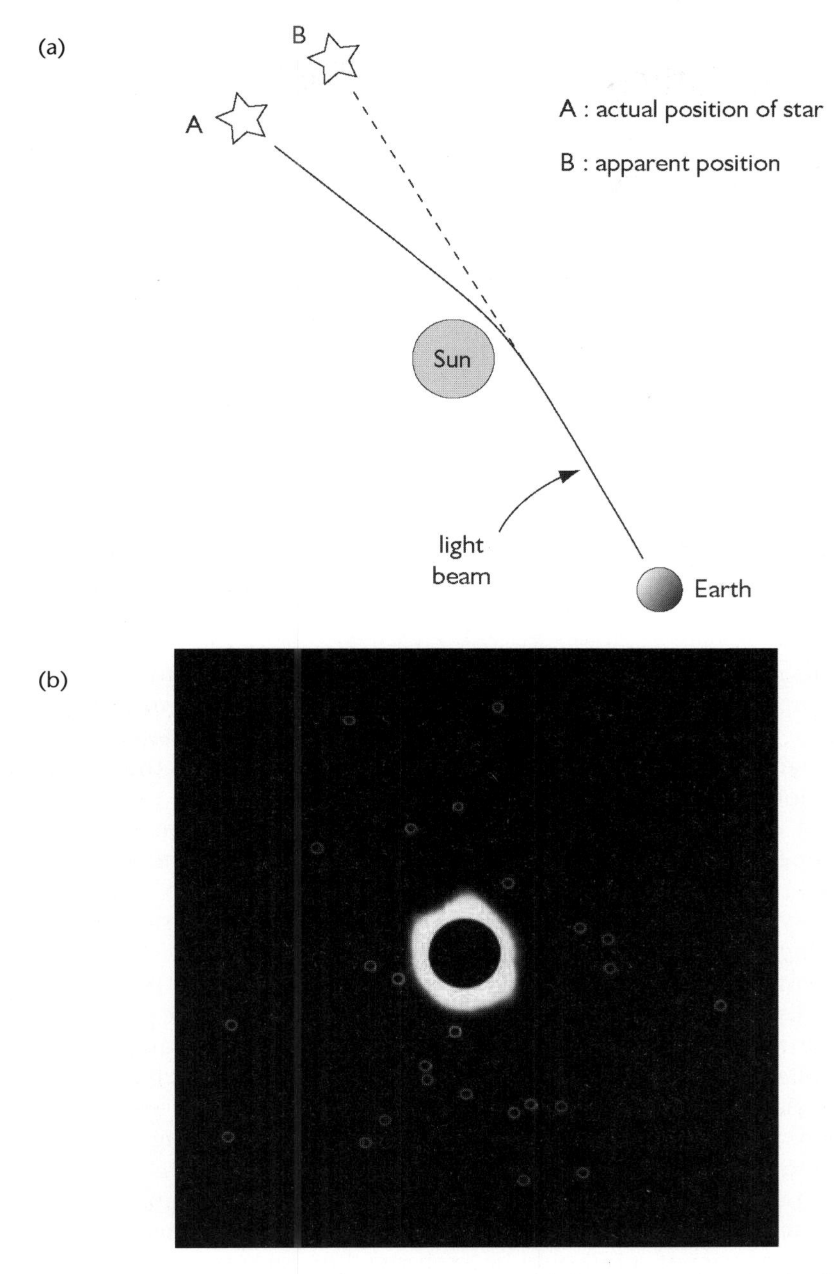


Figure 3.8 (a) The principle of the deviation of light passing close to the Sun (Royal Astronomical Society). (b) Photograph of the 1919 eclipse from the island of Principe, with positions of stars indicated. These positions are compared with a photograph of the same stars taken at night.

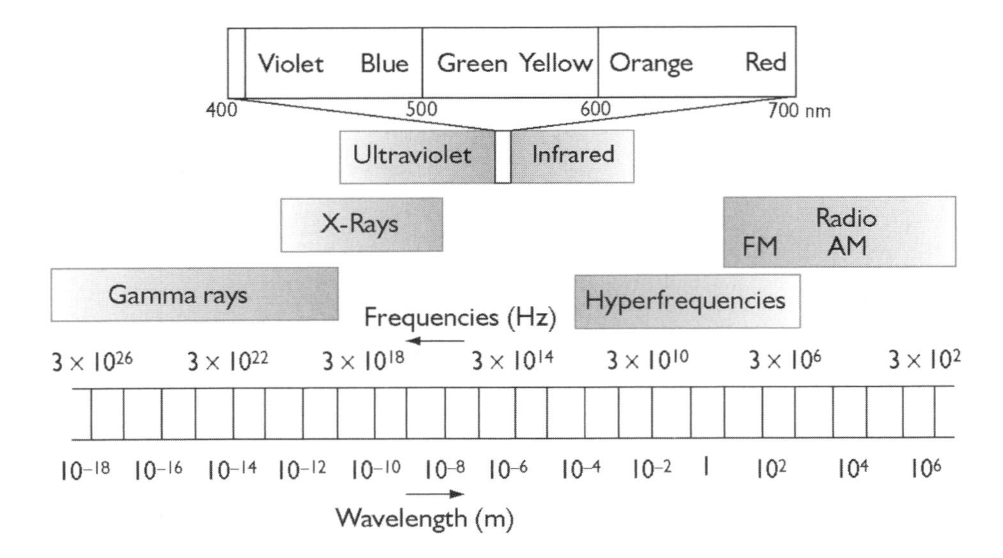


Figure 3.9 The electromagnetic spectrum. Visible light occupies a narrow band between 400 and 700 nanometres (billionths of a metre).

In order to model a radiating body without having to take into account its chemical composition, physicists consider an ideal radiator, known as a 'black body': within a heated enclosure, a thermal equilibrium will be established between the radiation emitted and the radiation absorbed. What is the spectrum of this radiation at equilibrium? Can we calculate the intensity of the radiation of the heated body at temperature $T$, as a function of its frequency? German physicist Wilhelm Wien had proposed a mathematical function which described this spectrum by way of two parameters. There were, however, two problems, in that no theory could explain it, and it did not tally with experimental data obtained at higher frequencies – towards the ultraviolet.

The German scientist, Max Planck, determined to resolve this problem and began with a hypothesis inspired by the kinetic theory of gases. The properties of a gas reflect the behaviour of its constituent atoms, which are colliding. Similarly, radiation from a black body may be considered as a collection of microscopic electromagnetic oscillators, emitting and absorbing radiation. In order to apply this model, Planck set out to determine the exchanges of energy between the 'oscillators', but was faced by a difficulty. If these exchanges were continuous, the energy given might be subdivided in an infinite number of ways. From this came Planck's second hypothesis, which he labelled an 'act of despair'. He supposed that energy could only be exchanged in packets, or quanta, with values proportional to the frequency of the emitting oscillator.

If this were proved, then energy exchanges could be represented with finite numbers, and the kinetic theory of gases could be applied. In 1900, Planck described in mathematical terms the spectrum of the black body. The constant of proportionality between frequencies and energy exchanges, *h*, is now known as Planck's constant. When it appeared, it was a modest category A constant, since it seemed to characterise only a particular physical mechanism: the energy exchanges between radiation and the black body. It was thought that it might be possible to derive from this a better understanding of the interaction between matter and radiation. Planck himself thought that this 'discretisation' of energy exchanges was but an artifice of calculation.

## QUANTA OF LIGHT

Once again, it was Einstein who realised the importance of Planck's discovery, interpreting it in another of his famous articles in 1905. This was to prove an *annus mirabilis* for the German physicist, not unlike Newton's fruitful work of 240 years before. The one-time patent-office clerk now had electromagnetism in his grasp. He understood that the relationship between energy and frequency of radiation was independent of the process of emission or absorption of that radiation. This relationship characterised radiation itself, and not the energy exchanges involved, as Planck had believed. With his characteristic dogged reasoning, Einstein proposed that there was a discontinuous distribution of

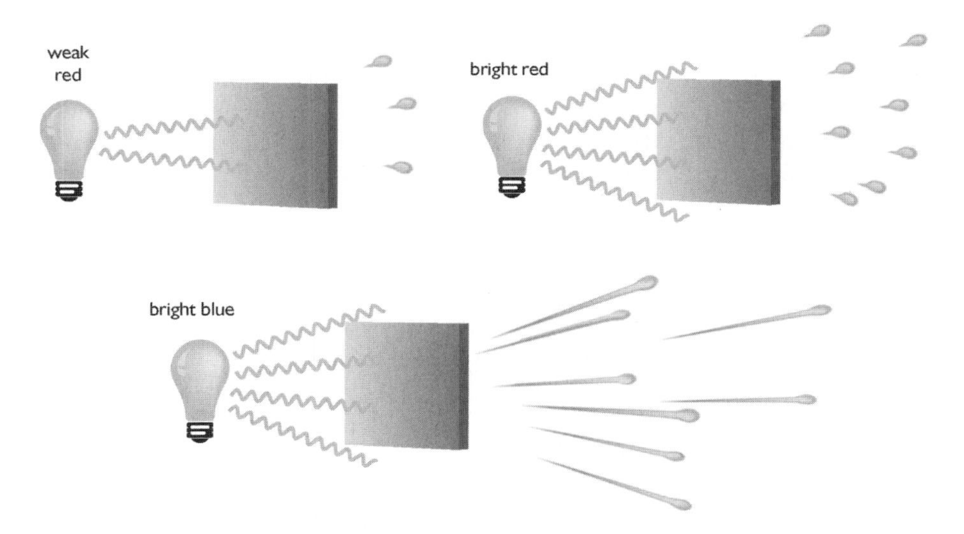


Figure 3.10 The photoelectric effect. A dim red light ejects a few electrons from a metal plate. A brighter light ejects more electrons, but these electrons do not propagate faster. However, a blue light (thus of higher frequency) causes the ejected electrons to propagate faster. This shows that the energy of the photons, converted into the kinetic energy of the electrons, is proportional to their frequency. (After *Pour la Science*, December 2004.)

radiation, in the form of 'light quanta'. The relationship of proportionality between energy and frequency of radiation is today designated by the names of both scientists: the Planck–Einstein relation. To underline the difference between Planck's and Einstein's analyses, we recall a humorous observation attributed to Einstein: although beer is sold in pint bottles, this does not imply that it only exists in indivisible quantities of one pint. According to Einstein, this indivisibility is true for light, i.e. light (in contrast to beer) exists only in indivisible pieces.

Einstein's light quanta proved useful to him in the explanation of a hitherto mysterious phenomenon: the photoelectric effect (Figure 3.10). Hertz (famous for his radio waves) had observed in 1887 that a metal block illuminated by a beam of light emits electrons. However, this effect happened only at high frequencies, and disappeared below a certain threshold frequency, irrespective of the intensity of the beam. Moreover, the speed of the ejected electrons depended on neither the frequency of the beam nor its intensity. This experimental fact can be explained using the idea of quanta. In effect, each quantum of light brings its energy to the block of metal, this energy being proportional to its frequency. If the frequency of the quantum is sufficiently high, it will be able to dislodge an electron from the block, and the electron will move away at a certain speed. If the frequency is too low, the quantum of light will not carry sufficient energy, and the electron will remain in its place in the metal block.

Einstein's article on the quanta of light and the photoelectric effect, which won him the Nobel Prize in 1921, showed that Planck's constant $h$ was involved in all luminous processes. Therefore, only five years after its inception, $h$ moved from category A into category B. In 1916, having studied the photoelectric effect in minute detail, Robert Millikan, an American scientist-experimenter who had already determined the charge of the electron some years earlier, measured $h$. His value was $6.6 \times 10^{-34}$ joule-seconds (J.s), in agreement with that determined by Planck from his black-body spectra. This value is extremely small because, even at the characteristic frequencies of visible light – of the order of $10^{15}$ hertz, or one million billion vibrations per second (the energy of a quantum is close to a billionth of a billionth of a joule). Nevertheless, this constant became the key to a whole world of atoms.

## A PLANETARY SYSTEM IN THE ATOM

Einstein used the photoelectric effect to show that quanta of light, later called *photons*, fitted a corpuscular description of light. Now that the wave nature of light had been solidly established through optics and electromagnetism, would it be necessary to turn back to the old-fashioned corpuscular theory? No! These two representations of light, as waves and as corpuscles, were to be reconciled. Planck's constant built a bridge between the two by linking quantities that hitherto were not commensurable: energy and frequency or, in other words, momentum and wavelength. A new mechanics would arise from this unification.

Firstly, quantum mechanics, as it was eventually called, had to explain why atomic emission spectra exhibited bright lines at precise frequencies (what Maxwell called their 'vibrations'), rather than being continuous 'rainbow' spectra. To understand this, it would be necessary to explain the structure of the atom itself. Danish physicist Neils Bohr conceived a 'planetary' model of the atom in 1913. Similar to the manner in which planets orbited the Sun owing to its gravitational effect, the (negatively charged) electrons would orbit the atomic nucleus, which kept them in their paths with the attractive force of its positive electric charge. In accordance with his 'quantification rule', Bohr introduced discontinuity into the system by having the electrons pursue only certain permitted orbits (Figure 3.11), assuming that the angular momentum of the electron – i.e. the vector product of its position and momentum – could be equal only to an integer multiple of the Planck constant. This quantification hypothesis led Bohr to a resolution of the question of the position of the bright spectral lines of hydrogen. He seemed to be on the right track.

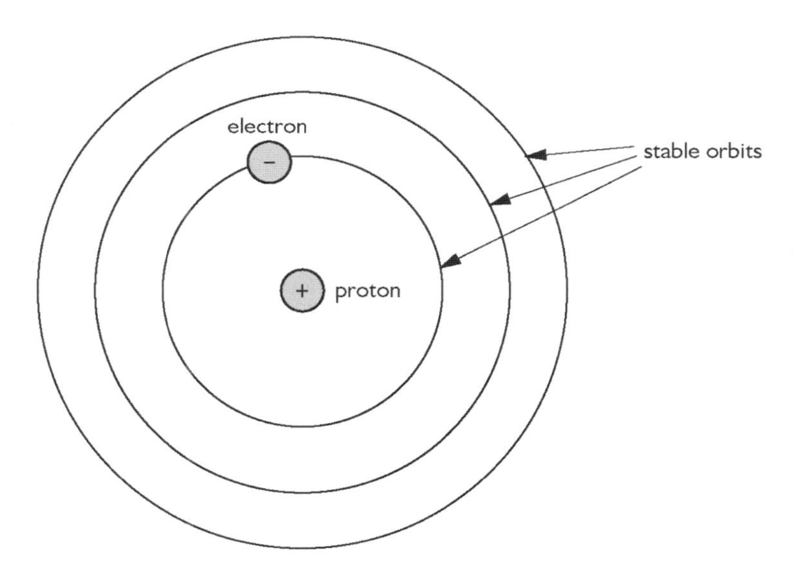


Figure 3.11 The Bohr atom. In the case of the hydrogen atom, the nucleus is composed of a single proton, with positive charge, around which 'orbits' an electron, with negative charge. Only certain orbits are permitted.

It now appeared that Planck's constant was a measure of angular momentum, a quantity defined by Newtonian mechanics. So the electron remained a 'corpuscle', endowed with a position and momentum. A young French physicist, Louis de Broglie, was not completely convinced by this: the fact that integers were involved in the behaviour of atomic electrons suggested to him that successive lines were the result of a wave-like phenomenon. He proposed that the Planck–Einstein relation should also be applied to electrons, in order to assign a wavelength (or frequency) to them. Since light, which was thought to travel as a wave, also showed evidence of a corpuscular nature, why should (corpuscular) electrons not also be waves?

This wave-like nature of electrons was verified experimentally in 1927: a crystal diffracted a beam of electrons, just as it diffracted a beam of X-rays. Henceforth, Planck's constant would be identified not only with light, but also with all matter. It had moved into our category C of universal constants.

## SOME DISCONCERTING PHYSICS

Quantum mechanics still held some surprises, and the role of the constant $h$ would reveal itself to be even more profound than had been imagined. A wave function – a concept transcending both particles and waves – was associated with electrons. On this basis, physicist Erwin Schrödinger established in 1925 his equation of temporal evolution. But what did this wave function represent? Bohr and his colleagues interpreted it as the probability of the presence of the quantum at any given point. The *probability* of its presence? The introduction of 'chance' into the laws of physics caused furrowed brows on more than one researcher, including Einstein. It had to be admitted that, before any measurement was made, the position of a quantum remained 'unspecified'. Similarly, the traditional idea of a trajectory was also held in suspicion, as the same 'indeterminacy' was associated with the momentum of the quantum.

Was it therefore not possible to describe the microscopic world? In Germany, the physicist Werner Heisenberg confirmed this when he announced his Uncertainty Principle. The product of uncertainties in simultaneous measurements of position and momentum must exceed the value of Planck's constant divided by $4\pi$. Of course, given the tiny value of constant $h$, this indeterminacy remains minuscule, its value often less than that of experimental uncertainty. However, it does indicate a characteristic of the behaviour of Nature. Just as $c$ and $G$ summarise the properties of space-time and space-time-matter, $h$ imposes limits on the investigation of any physical system.

## SYNTHESES AND LIMITS

Our inquiry into physics has shed light on the nature of fundamental constants. Figure 3.12 summarises the progress of the most fundamental among them, namely $c$, $G$ and $h$, giving dates for their transitions from one category to another, all the way to their status today. While relating their histories, we have noticed some of their peculiarities.

Firstly, their status is not immutable, and, as new discoveries are made, a constant may pass from one category to another, and may even run the risk of losing its status as a fundamental constant. We may therefore have reason to doubt the immutability of constants.

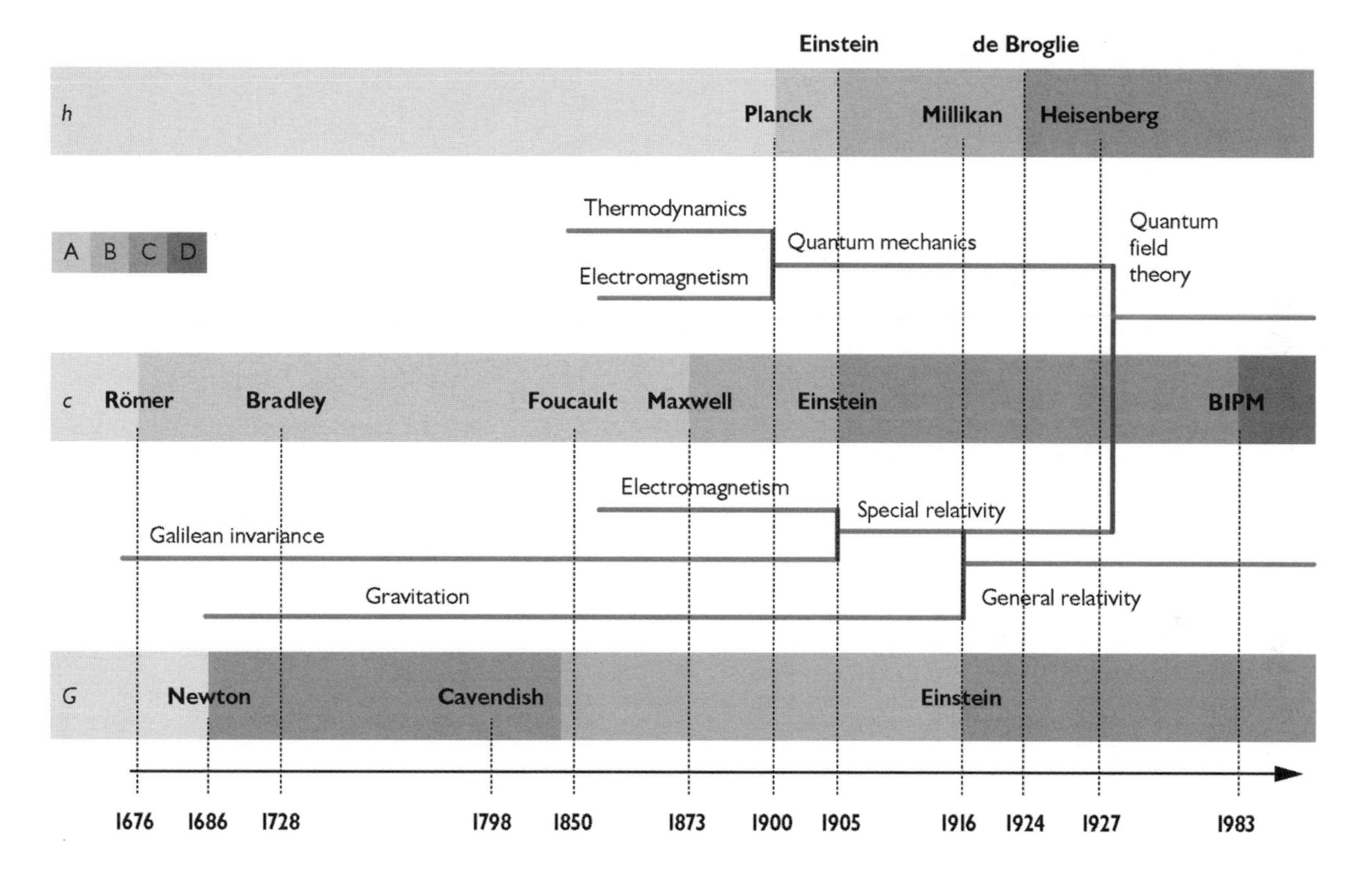


Figure 3.12 Timelines of the evolution of the status of *c*, *G* and *h*, compared with the advent of new theories (for example, the inconsistencies between thermodynamics and electromagnetism led to the formulation of quantum mechanics).

Secondly, their acceptance or changes of status are associated with major theoretical revolutions in physics. Once contradictions between existing theories have been resolved, the way is open for new concepts that are more general or more synthetic than the old. Constants build bridges between quantities hitherto considered to be incommensurable, and allow new concepts to emerge. Thus, for example, the speed of light underpins the synthesis of space and time, and the gravitational constant links matter to space-time. Planck's constant involves the relationship between *energy*, a quantity linked to the corpuscular, and *frequency*, a quantity associated with waves. All these constants led to a unification.

Thirdly, fundamental constants enable us to define the domains or the validity of these theories. The speed of light appears to be a universal limit. As soon as a physical phenomenon involves speeds approaching $c$, relativistic effects become important; if, on the other hand, speeds remain well below that of light, relativistic effects are negligible, and Galileo's mechanics will suffice. Planck's constant also acts as a referent, since, if the 'action' of a system greatly exceeds the value of that constant, classical mechanics will give the appropriate answers. But if this action becomes comparable to Planck's constant, quantum effects (such as the corpuscular behaviour of light or the wave behaviour of matter) will intervene. The case of the gravitational constant is less clear, since no physical quantity possesses its dimensions.

## THE CUBE OF PHYSICAL THEORIES

We shall summarise the roles of $c$, $G$ and $h$ in the structuration of physical theories by allowing ourselves a little geometrical diversion. We are going to construct a 'cube of physical theories' in which each of the eight corners represents a theory (Figure 3.13). We create our cube by first tracing out three perpendicular axes, one for each constant: $G$, $h$ and $1/c$. On each axis, the associated constant can be zero or non-zero, where 'zero' signifies its 'non-existence'. If $G$ is zero, no gravitational effect is considered; if $1/c$ is zero, it is assumed that the speed of light is infinite, as was long believed in Newtonian mechanics; and if $h$ is zero, the quantum behaviour of light and matter is not taken into account. So, at the point of origin of our axes (corner 1), we find classical mechanics.

Each time we 'switch on' a constant, new physical phenomena appear, which are characteristic of theories at that level. Let us give each of our constants in turn a non-zero value, leading us successively towards the corners lying on the axes – that is, corners 2, 3 and 4. At corner 2, on axis $G$, we find the Newtonian theory of gravitation; no relativistic or quantum effects yet intervene. Corner 3, at which $c$ is finite, denotes special relativity, unifying space and time; this theory does not involve gravity or quantum phenomena. On the $h$ axis lies quantum mechanics (corner 4), which associates the corpuscular and wave characteristics of light and matter, but does not take into account relativistic or gravitational effects.

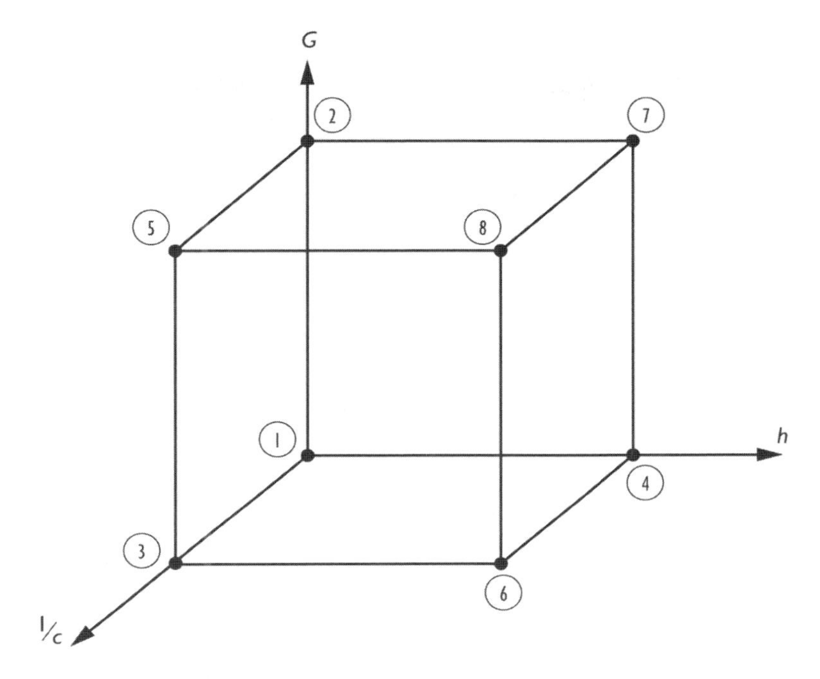


Figure 3.13 The cube of physical theories, based on the three constants $c$ (or rather $1/c$, which is zero when $c$ equals infinity), $G$ and $h$. On the cube are: classical mechanics (1), Newtonian gravitation (2), special relativity (3), quantum mechanics (4), general relativity (5), quantum field theory (6), a quantum theory of gravitation which does not yet exist (7) and a 'theory of everything', the dream of some physicists (8).

All three of these theories have been shown to be very effective in their own authenticated fields.

Other theories bring in two constants at once, which takes us to corners 5, 6 and 7. Through the use of constants $c$ and $G$, general relativity (corner 5) leads to the unification of space-time and matter. At corner 6, quantum field theory deals with quantum and relativistic effects. It offers a unified description of three interactions in nature: electromagnetic, strong and weak forces. General relativity and field theory are well established both theoretically and experimentally. On the other hand, at corner 7, there does not yet exist a coherent, non-relativistic quantum theory of gravitation that simultaneously involves $G$ and $h$.

Corner 8 of our cube, which we create by using a non-zero value for all three constants, would represent the 'theory of everything', as envisaged by some physicists. It would offer a description of gravity which is at once quantum and relativistic. String theories, so intensively studied nowadays, are serious candidates for this eighth corner. As these theories have many implications for constants, read on ...

Contrary to what our commission of inquiry might have feared, a possible variation of constants would not threaten the findings and the integrity of physics; it would rather increase our hopes that the edifice might be completed. Observing the variation of a constant could point physicists along a path towards that grand unified theory they so fervently seek. We still require to know the constants that may vary and the observable effects that any such variations would have.

# 4

# Proportions: dimensionless parameters

So far, our inquiry shows that physics rests on solid foundations, its structure being supported by the fundamental constants. However, we have as yet said nothing about the possible *variation* of a fundamental constant. Which constants could vary? And in what proportion? What observable effects would there be?

To find the numerical value of fundamental constants, we have to resort to the process of measurement. The numerical value of a variable or a constant depends closely on the system of units used to measure it. Conversely, the value of a constant can determine a unit: this is the case with the speed of light, which is used to define the metre. Before we can discuss the values of constants, we must first unravel the ties that connect them to units.

## COHERENT UNITS

When we consulted our dictionary for the definition of the word 'constant', we found the expression 'universal physical constant': a "quantity which, measured within a coherent system of units, is invariable". But what is a coherent system of units? It is a system that associates a precise unit with each dimension. For example, in the case of Newton's force of gravity, the masses are expressed in kilogrammes, the lengths in metres, and the times in seconds. We are therefore obliged to express the gravitational constant with some compatible unit – for example, cubic metres per kilogramme per second squared ($m^3/kg/s^2$). If these units are used, $G$ has the value $6.67 \times 10^{-11}$.

We could, of course, as has already been noted, decide to use tonnes for our masses, millimetres for lengths and minutes for times, in which case $G$ would assume the value 185 ($mm^3/t/min^2$). These units might well seem to be more practical, but they could be leading us astray, because metres and kilogrammes 'hide' within other units. Let us consider the strength of a force: it is measured in newtons. Now, we know (see Chapter 1) that the force transmits an acceleration to a mass, such that it possesses the dimensions of a mass multiplied by a distance and divided by the square of a time ($ML/T^2$). Consequently, newtons are merely kilogramme-metres per second squared ($kg.m/s^2$). What happens if, in some physical description, a force is expressed in newtons while a mass is expressed in tonnes? In such a combination we get an error factor of 1000.

To show how important it is to get the units right in numerical applications, let us recall the unfortunate example of the *Mars Climate Orbiter* spacecraft. In September 1999, the attempt to insert this probe into an orbit around the Red Planet failed lamentably, even though its voyage had, up to that point, been problem free. What went wrong? Two international teams, within NASA, had the job of overseeing the approach manoeuvres of the orbiter. Unfortunately, one team was working in imperial units (miles and pounds), while the other was using SI (metric) units. Nobody realised this until it was too late, with the result that the probe missed its intended target by 100 kilometres (just over 60 miles) and crashed on Mars. We can only hope that something was learned from this misadventure. . .

## THE *SYSTÈME INTERNATIONAL (SI)*

The choice of a coherent system of units is vital in physics. Nowadays, we use the international system (SI) of units, which has its roots in the metric system that was defined during the French Revolution. This system has evolved with the flow of new discoveries and methods.

One of the first to employ a coherent system of units in his own work was the German physicist and mathematician Carl Friedrich Gauss. In the first half of the nineteenth century, in his observatory in Göttingen, Gauss carried out the first detailed study of the Earth's magnetic field. He wanted his data to be accessible to the greatest possible number of physicists, and to this end he used the new French units. Gauss expressed lengths in millimetres, masses in grammes and times in seconds. He expressed values for the Earth's magnetic field in a unit of magnetic flux, the weber, named after his collaborator Wilhelm Weber. The Earth's magnetic field is known to create a magnetic flux of approximately half a weber.

In 1873, James Clerk Maxwell and William Thomson (the latter recently elevated to the title of Lord Kelvin) emulated Gauss and Weber, defining and promoting the *CGS system*. As the abbreviation indicates, lengths in this system are measured in centimetres, masses in grammes, and times in seconds. The weber was replaced by a unit of magnetic flux density, the gauss, with a value of 1 weber per square centimetre. All the other dimensions were based on these four units: surfaces were measured in square centimetres, volumes in cubic centimetres, forces in dynes (1 dyne is the force that accelerates a mass of 1 gramme at a rate of 1 centimetre per second squared), etc. At the same time, a parallel system was being used, the MKS system, which was based directly on the 'prototypes': metres, kilogrammes and seconds.

Now the study of electromagnetism was proceeding in leaps and bounds, and many physicists felt the need for units of measurement that were specific to this field – enough to facilitate research, but not so many as to create redundancy in the system. For example, using both the gauss and the weber within the same system of units would have been unnecessary and superfluous. At the beginning of the twentieth century, Italian physicist Giovanni Giorgi demonstrated that a coherent system could be created using only four base units. As the gauss had been shown to be an impractical unit for the description of electromagnetic phenomena, it was replaced by the ampere, a unit of electric current, giving rise to the so-called MKSA system.

There were other fields in which physicists felt that new units were needed. In 1954, the kelvin (1/273.16 of the temperature of the triple point of water) was chosen for the measurement of temperature (0 °C therefore corresponds to 273 kelvins, the kelvin scale starting from absolute zero). In the same year, a new unit of luminous intensity, the candela, entered the system and in 1966 the mole (the amount of matter in 12 grammes of carbon-12) was introduced.

Today's *Système International* has seven base units: the metre, the kilogramme, the second, the ampere, the kelvin, the mole and the candela (Figure 4.1). From these seven, other units can be expressed directly: for example, square metres for surfaces, cubic metres for volumes and metres per second for speeds. Some units derived from the basic seven can also have names of their own, such as the newton for a force (metre-kilogramme per second squared) and the joule for energy (newton multiplied by metre) (Figure 4.2).

| | | |
|---|---|---|
| length | metre | m |
| mass | kilogramme | kg |
| time | second | s |
| electric current | ampere | A |
| thermodynamic temperature | kelvin | K |
| intensity of light | candela | cd |
| amount of matter | mole | mol |

Figure 4.1 The seven base units of the *Système International* (SI). (After CIPM.)

## A TRIO OF IRREDUCIBLES

In the final analysis, how many units are required? As Giorgi pointed out, physicists can derive all the other units they need from only four of the seven independent units. So the other three SI units would seem to be redundant, but they reflect a compromise between experimental pragmatism and the idealism of theoreticians.

To verify Giorgi's statement, let us examine how the kelvin, the mole and the candela are derived from the metre, the kilogramme, the second and the ampere. We know from the kinetic theory of gases that temperature is a measure of the internal motions of the constituents of matter, and can therefore be described in terms of their mean kinetic energy and expressed though mechanical units of

| | | |
|---|---|---|
| area | square metre | $m^2$ |
| volume | cubic metre | $m^3$ |
| frequency | hertz | Hz = 1/s |
| density | kilogramme per cubic metre | $kg/m^3$ |
| velocity | metre per second | m/s |
| angular velocity | radian per second | rad/s |
| acceleration | metre per second squared | $m/s^2$ |
| angular acceleration | radian per second squared | $rad/s^2$ |
| force | newton | $N = kg \cdot m/s^2$ |
| pressure (mechanical tension) | newton per square metre | $N/m^2$ |
| kinematic viscosity | square metre per second | $m^2/s$ |
| dynamic viscosity | newton-second per square metre | $N \cdot s/m^2$ |
| work, energy, heat | joule | $J = N \cdot m$ |
| power | watt | W = J/s |
| charge | coulomb | $C = A \cdot s$ |
| electrical potential, electromotive force | volt | V = W/A |
| electric field intensity | volt per metre | V/m |
| resistance | ohm | Ω = V/A |
| capacitance | farad | $F = A \cdot s/V$ |
| magnetic flux | weber | $Wb = V \cdot s$ |
| inductance | henry | $H = V \cdot s/A$ |
| | | (conductance); |
| | | magnetic |
| induction | tesla | $T = Wb/m^2$ |
| magnetic field intensity | ampere per metre | A/m |
| magnetomotive force | ampere | A |
| luminous flux | lumen | $lm = cd \cdot sr$ |
| luminance | candela per square metre | $cd/m^2$ |
| illuminance | lux | $lx = lm/m^2$ |

Figure 4.2 Derived SI units. (After CIPM.)

length, mass and time. As for the mole, it is merely a unit denoting numbers of molecules (1 mole = 6 × $10^{23}$ molecules, which is Avogadro's number, and is another physical constant!). As the candela is expressed in terms of energy flux, it is also a mechanical quantity.

We are therefore, *a priori,* in agreement with Giorgi. Four units are ample, but are they all necessary? Could we perhaps produce a coherent system from only three base units? In reality, as Gauss remarked, the ampere can be derived from the three mechanical units: note that the intensity of an electric current represents a flow of electric charges, i.e. a quantity of charges per unit of time. Now, what is the unit of electric charges? If we can express it in mechanical units, we can do the same with the ampere. The history of Coulomb's law is worth a brief mention in this context.

In about 1780, Charles Augustin de Coulomb constructed a torsion balance in order to study electrical forces. With this apparatus, he was the first person to measure electric charges, and he did it in the 'traditional' manner of measuring them against each other. Contact with a charged sphere would create other spheres carrying, first, a quarter, then an eighth, and so on, of the original charge. He showed that the force between two electric charges is proportional to the product of the charges and inversely proportional to the square of the distance between them. Coulomb's law led physicists to define a new unit to designate the electric charge, and, for obvious reasons, it was called the coulomb (abbreviation: C). This law also meant that a new constant, the dielectric constant, came into being. In mathematical terms, the coulomb force between two charges $q_1$ and $q_2$ is written $F = \kappa q_1 q_2/r^2$, where the dielectric constant $\kappa$ has the dimensions of a force multiplied by a distance squared and divided by a charge squared. In the *Système International,* $\kappa$ is measured in newton-square metres per coulomb squared ($N.m^2/C^2$).

In summary, Coulomb's law implies a new unit, the coulomb, and a new constant, the dielectric constant. However, we could have used a different approach: instead of introducing constant $\kappa$, we could have incorporated it into the definition of the electric charge $q$, and could simply have written $F = q_1 q_2/r^2$. In this case, the unit of charge is derived directly from the mechanical units, the newton and the metre. Consequently, the same would be true of the ampere (electric current). This very economical approach does not require any new constant or unit to describe electrical phenomena. So, having established this, we see that the total number of units necessary in physics is limited to three: those that measure lengths, masses and times.

## DIMENSIONLESS PHYSICS?

Why stop there, when we are doing so well? After all, we could continue this process of the reduction of the number of units and finally eliminate them all. As soon as a constant links two quantities that were hitherto considered independent, we can consider it to be simply a conversion factor between them.

This is what happened to Joule's constant, which linked work and heat. When it was understood that heat could be measured in units of energy, the calorie and Joule's constant both became obsolete at a stroke.

In 2003, British physicist Mike Duff pursued a similar line, reducing to one the three dimensions of physics (length, mass and time). But how did he achieve that?

First, it is easy to reduce lengths to times by bringing in the light-speed constant $c$ as a conversion factor between the two quantities: we make the numerical value of $c$ equal to 1, and measure lengths in light-seconds. Now only two dimensions remain: time and mass.

Then, recalling that the Planck–Einstein relation links energy to frequency (i.e. the inverse of time), we can measure time as the inverse of energy. In 1905 Einstein showed that energy equals mass. Consequently, time can be measured as the inverse of mass. All that is needed is to count the number of time periods on a quantum oscillator of energy equal to 1 kilogramme.

So, by assigning a numerical value of 1 to $c$ and $h$, we have reduced all the fundamental dimensions to only one: mass. At this stage, we can still choose a new mass standard such that the numerical value of $G$ equals 1. As we have already emphasised, it is not so much the numerical values of constants that matter, but the bridges that these constants make between different concepts. Our choice consists in selecting a particular system of units, defined by the fact that the numerical values of our three constants all equal 1. Such a standard exists: at least in theory. It is a microscopic black hole of radius equal to its quantum wavelength. This is surely one unusual 'prototype' that the scientists at the International Bureau of Weights and Measures (BIPM) are not yet prepared to keep in their metal cabinet in the Pavillon de Sèvres!

## NATURAL UNITS

Joking apart... we have obtained, by following Duff's reasoning, a system within which the numerical values of $c$, $h$ and $G$ all equal unity. The idea of using these three fundamental constants to create a system of units had already been suggested by Planck. He thought it was only necessary to combine these three fundamental constants to define standards for length, mass and time. The Planck length ($l_P$, the letter P referring to Planck), is defined from these three constants. It is written $l_P = \sqrt{(Gh/c^3)}$ and has a value of $4.04 \times 10^{-35}$ metres. The Planck mass, similarly expressed, is $m_P = \sqrt{(hc/G)}$ with a value of $5.45 \times 10^{-8}$ kilogrammes. This corresponds to the mass of the black hole proposed by Duff. The Planck time is $t_P = \sqrt{(Gh/c^5)}$ with a value of $1.35 \times 10^{-43}$ seconds. The proposition, therefore, was to utilise Planck's system as a system of units.

Twenty-five years before Planck, another scientist had suggested a similar system of units based on another system of constants. The brilliant Irish physicist George Johnstone Stoney is remembered as the 'inventor' of the electron. In 1874, he supposed that the electric current consisted of 'grains' of elementary charges, which he called 'electrons' in 1891. A British physicist, Joseph John Thomson, isolated the electron and identified it in 1897. Meanwhile, Faraday had achieved the electrolysis of water, wherein molecules become positively charged ions after losing one electron. Having carried out his own electrolytic experiments, Stoney converted his value for the elementary charge into our *Système International*, and found it to be (approximately) $10^{-20}$ coulombs, which is not far from the modern value of $1.6 \times 10^{-19}$ coulombs. After formulating his hypothesis, Stoney then formulated a system of units based on the gravitational constant $G$, the speed of light in a vacuum $c$ and the elementary charge, known as $e$. In our *Système International*, the Stoney length is expressed by $l_S = \sqrt{(Ge^2/4\pi\varepsilon_0 c^4)}$, giving a value of $1.37 \times 10^{-36}$ metres. The factor $4\pi\varepsilon_0$ has its origin in the dielectric constant, and is involved in the definition of the electric charge which figures in the *Système International*. The Stoney mass is expressed by $m_S = \sqrt{(e^2/4\pi\varepsilon_0 G)}$, giving a value of $1.85 \times 10^{-9}$ kilogrammes; and the Stoney time is expressed by $t_S = \sqrt{(Ge^2/4\pi\varepsilon_0 c^6)}$, with a value of $4.59 \times 10^{-45}$ seconds.

Stoney's system is based on the use of $c$, $G$ and $e$ as fundamental units, while Planck's system is based on $c$, $G$ and $h$. There are other possibilities. For example, we can amuse ourselves by creating a system of units based on the mass of the electron, the electric charge and Planck's constant ($m_e$, $e$ and $h$). We could call this the 'Bohr system', since it is suitable for the study of the atom, which so preoccupied Niels Bohr. We thus obtain the Bohr length, $l_B = h^2 4\pi\varepsilon_0/m_e e^4$, with a value of $2.1 \times 10^{-9}$ metres; the Bohr mass, which is given directly by $m_e$ and has a value of $9.1 \times 10^{-31}$ kilogrammes; and the Bohr time, $t_B = h^3(4\pi\varepsilon_0)^2/m_e e^4$, with a value of $6.0 \times 10^{-15}$ seconds.

We see that the choice of constants defining these three systems of units is arbitrary, but, as our analysis of independent dimensions has revealed, we will always require three constants of independent dimensions to define a system of units.

These three systems of units all obey Maxwell's call for universality and objectivity. There is no reference to humankind or the Earth. Planck even mentioned the idea of a universal system "independent of any specific substance, holding its meaning for all time and with all civilisations, including any who are not of this human world". Such a solution would satisfy the theoreticians' idealism. However, the practical use of these systems of units remains limited, for two reasons. Firstly, the standards obtained in this way are abstract and almost impossible to realise in a laboratory; and, secondly, their minuscule values are too far removed from the needs of experimentalists.

It seems therefore preferable, from the practical point of view, to retain three fundamental constants whose values, determined through experiment, are expressed in terms of the three fundamental dimensions of physics, and the international system of units.

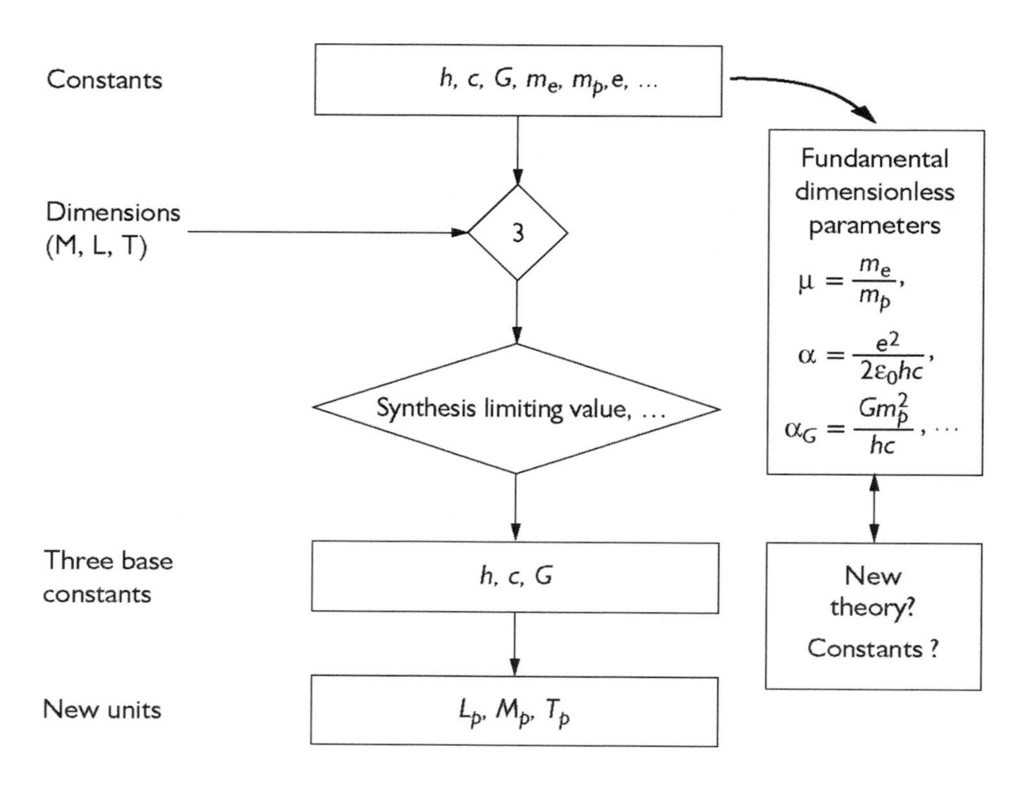


Figure 4.3 Three fundamental units... three fundamental constants. All the units may be derived from three of them (mass, length, time). Similarly, we select three of the physical constants ($c$, $G$, $h$) for their properties of universality, synthesis and limiting value. By combining these three fundamental constants, we define new units, in this case Planck's system. By expressing the other constants in this system of units, we obtain dimensionless parameters, characterizing our physical world.

## ESSENTIAL PARAMETERS

Nevertheless, there is one case where the use of the three fundamental constants as standard measures proves of great use and significance: in the study of physical constants. From the list of constants drawn up in Chapter 2, we can still isolate three (dimensional) constants that, by comparison, are the most fundamental, and can be used to express the functions of all other constants (Figure 4.3).

Let us consider an example. We shall choose the mass of the electron as one of our three fundamental dimensional constants. All other masses in the list of constants are expressed as a function of this one. For example, the mass of the proton is 1836 times that of the electron. This is a dimensionless number, which means that it retains this value in all systems of units. Similarly, the mass of the neutron and all constants of mass, expressed with reference to that of the electron, assume dimensionless values, independent of the system of units.

Proceeding thus, we obtain ratios of the magnitudes of masses, charges and forces, etc., that no definition or change in units will be able to modify. In this way we discover the relationship between scales. Like Dirac, we confirm that the ratio of the strengths of gravitational and electric forces between the proton and the electron equals $10^{39}$, irrespective of the system of units.

From a more complex combination of fundamental constants, we infer another dimensionless parameter, of great importance in physics: $\alpha = e^2/2\varepsilon_0 hc$. In this equation, $\alpha$ is the fine-structure constant, the same one that led us to undertake our inquiry into the constancy of constants. Through the intermediary of the factor $e^2$ in the expression, $\alpha$ expresses the strength of the electromagnetic coupling between charged particles. Its value, remember, is approximately 1/137.

That the proton should be 1836 times heavier than the electron is not an insignificant fact: this figure has its importance in the equilibrium of atoms. The mighty relationship between gravitational and electric forces governs the dynamic of the many levels of Nature. As for the parameter $\alpha$, it is involved in the stability of atoms and chemical bonds: in other words, dimensionless parameters, derived from comparisons with dimensional constants, represent the very essence of Nature. To change their value would be to change the phenomena of Nature. In a certain way, the values of these parameters are of more importance than those of dimensional constants.

Albert Einstein expressed this point of view clearly in a letter to his friend and fellow physicist Ilse Rosenthal-Schneider, who had asked him about the significance of fundamental constants. Einstein wrote that there are two kinds of constants, *apparent* and *real*. Apparent constants result simply from the introduction of arbitrary units, but can be eliminated (by an arbitrary choice of a system of units). Real constants, claimed Einstein, are authentic numbers, which God must have chosen arbitrarily when he saw fit to create the world. In a later letter, he added that these (dimensionless) constants must be basic numbers whose values are established by the logical foundations of all theory. Einstein stated that he could not imagine a unified, reasonable theory that explicitly contained a number which the Creator, on a whim, would have chosen differently, and from which a qualitatively different set of laws governing the world would have resulted.

In other words, these dimensionless parameters are more likely to be the ones that a more fundamental theory of Nature might explain.

## WHAT IF THEY WERE DIFFERENT?

To illustrate the role of dimensionless parameters in the coherence of physics, let us try to imagine that some whim of the Creator has modified one of the parameters that characterise our world. George Gamow, a talented Russian physicist born in Odessa, came up with a famous example, in 1940, of this kind of mental exercise, in a popular science book entitled *Mr Tompkins in Wonderland* (Figure 4.4). Its hero, Mr C.G.H. Tompkins (do his initials remind us of anything?), visits places where the fundamental constants of physics differ markedly from those of our world. This ploy allows Gamow to transpose onto our scale phenomena that seem very strange to us, for example, quantum and relativistic effects. Mr Tompkins goes to a town where the speed of light is only 20 kilometres per hour, causing him to notice various relativistic effects:

Figure 4.4 Mr Tompkins witnesses the contractions in length predicted by special relativity. (© Dunod.)

objects and people moving uniformly relative to him appear flattened in the direction of their motion. This contraction of length is a consequence of Lorentz transformations.

If we possessed supernormal powers, would we be able to modify the speed of light without affecting any other physical property? To investigate this, let us adopt a natural system of units that does not depend on the speed of light and will therefore be unaffected by a modification of this constant. Let us chose, for example, the one we have labelled the 'Bohr system', founded upon the electric charge, the mass of the electron and Planck's constant. These three fundamental constants will remain fixed when the speed of light varies. In this system, the standard of speed is the Bohr speed, i.e. the velocity of the electron on its innermost orbit around the atom: $v_B = e^2/4\pi\varepsilon_0 h$. In order to alter the value of the speed of light without changing the laws of physics, we must modify the dimensionless ratio $c/v_B = 4\pi\varepsilon_0 hc/e^2$.

We recognise within this formula the inverse (apart from a factor of 2) of the fine-structure constant $\alpha = e^2/2\pi\varepsilon_0 hc$. This is therefore the dimensionless parameter that varies if $c$ is modified, assuming the electron charge and Planck's constant to be fixed. Now this parameter characterises the coupling of charged particles, which participates in many important phenomena. It is this coupling, for example, that keeps electrons in the vicinity of the nucleus of the atom, and any modification of the strength of this force would affect the cohesion of atoms. A decrease of the speed of light would therefore increase the electromagnetic coupling, rendering exchanges of electrons between atoms much more difficult. Chemical bonds would therefore be less firm, and much rarer. Chemistry would be changed, and with it, biology, with repercussions for the appearance of life itself, resulting as it does from assemblages of long molecules: DNA and proteins. It is therefore improbable that a real live Mr Tompkins could exist to ride his bike in a world where the speed of light is much less than it is in ours.

Dirac, too, played with the idea of the Creator allowing a constant – the gravitational constant – to vary. As he was working in atomic units, he manipulated another dimensionless number, $Gm_p^2/hc$, where $m_p$ is the mass of the proton. Here too, it is not just a question of changing units, but of defining the relevant dimensionless parameter.

## PRIME SUSPECT

The time has come to take stock of what we have learned during our exploration so far: certain facts have emerged concerning fundamental constants. First of all, the whole of physics, and therefore the constants that appear in its laws, aims at universality. These laws and constants are valid in the widest possible domain, as a result of the very way in which theories are constructed, as French mathematician and physicist Henri Poincaré pointed out:

> "There is not one law of which we can state with certainty that it was always as true in the past, as it is today. What is more, we cannot be certain that it can never be shown in the future that it was different in the past. And nevertheless, there is nothing in this to prevent the scientist from keeping faith with the principle of immutability, since no law will ever be relegated to the rank of a transitory law, without being replaced by another more general and more comprehensive law. Its fall from grace will be due to the advent of this new law, so that there will have been no interregnum, and the *principles* will remain safe. By these will the changes be made, and these same revolutions will appear to be a striking confirmation of them."

Poincaré's declaration perfectly sums up what we have learned about the evolution of laws and fundamental constants in physics.

We have also stated that constants that are thought today to be the most fundamental, play a structural role in current theories. They set the limits on the domains wherein phenomena involved in various theories appear. They preside in the successive unifications that aim to describe the greatest number of phenomena with the minimum number of laws. The bringing together of three fundamental constants would lead to a "theory of everything", the goal of many a physicist.

However, no theory predicts the value of its fundamental constants, so we are reduced to determining them by measurement. Their numerical value depends consequently on a system of units. We can reduce the number of dimensional constants to three, and express the other constants by comparing them with these three as reference standards. We therefore obtain a list of dimensionless parameters that characterise the relative amplitude of physical phenomena. It is these dimensionless parameters that categorise phenomena, revealing the proportions of Nature and making the world what it is. For this reason, theoreticians seek to explain their values while experimental scientists, in their search for clues to a more fundamental theory, verify these constants with ever-increasing accuracy.

Among these dimensionless parameters, the constant $\alpha$ is the object of continued attention. It has already been suspected, notably by John Webb and his colleagues, of manifest inconstancy, and our commission of inquiry must now give $\alpha$ its full attention.

# 5

# Atoms under close examination

We have noted the relevance of *dimensionless* parameters, obtained through the combination of fundamental dimensional constants, and only accessible to measurement independently of a system of units. Let us now focus upon one of them, the fine-structure constant $\alpha$. This parameter is written as $\alpha = e^2/2\varepsilon_0 hc$, incorporating the square of the electric charge and thereby characterising the amplitude of electromagnetic interaction. Its value, as we have already seen, is 1/137.03599976.

In the laboratory, physicists have determined the value of $\alpha$ with a high precision to 12 significant figures. Given the extreme accuracy of the value of $\alpha$, it was considered that it would be easy to detect any possible variation in this parameter. This explains the great proliferation of studies in this area, which we will explore in this chapter and the next. The physicists have not, however, come to a conclusion about this, since $\alpha$ has something up its sleeve. In order to search for any possible variation in the value of the fine-structure constant – that is, within the limits of what can actually be measured – they have had to perform some amazing feats of ingenuity. Spectroscopy, atomic clocks and nuclear physics have all been brought into play to confirm the constancy of $\alpha$.

## THE LINES EXPLAINED

Following the work of Neils Bohr, as mentioned above, quantum mechanics revealed the mechanism by which atoms emit spectral lines at very precise frequencies. We know that the atom consists of a nucleus surrounded by electrons. The negatively charged electrons ($-e$) are attracted by the positively charged protons ($+e$) in the nucleus. In a neutral atom, there are as many protons as there are electrons, and their number, known as the *atomic number*, constitutes their identity. The chemical elements are classified, in order of increasing atomic number, in the Periodic Table (also known as the Mendeleev table, after its creator – see Figure 5.1).

The element occupying the first box of this table, hydrogen, is the simplest of the atoms. It has just one proton and one electron. It is this atom that Bohr first modelled. Since the electron is 1836 times lighter than the proton, Bohr assumed that the proton was immobile, while the electron moved around it. The electron

| $^{1}$H | | | | | | | | | | | | | | | | | $^{2}$He |
|---|---|---|---|---|---|---|---|---|---|---|---|---|---|---|---|---|---|
| $^{3}$Li | $^{4}$Be | | | | | | | | | | | $^{5}$B | $^{6}$C | $^{7}$N | $^{8}$O | $^{9}$F | $^{10}$Ne |
| $^{11}$Na | $^{12}$Mg | | | | | | | | | | | $^{13}$Al | $^{14}$Si | $^{15}$P | $^{16}$S | $^{17}$Cl | $^{18}$Ar |
| $^{19}$K | $^{20}$Ca | $^{21}$Sc | $^{22}$Ti | $^{23}$V | $^{24}$Cr | $^{25}$Mn | $^{26}$Fe | $^{27}$Co | $^{28}$Ni | $^{29}$Cu | $^{30}$Zn | $^{31}$Ga | $^{32}$Ge | $^{33}$As | $^{34}$Se | $^{35}$Br | $^{36}$Kr |
| $^{37}$Rb | $^{38}$Sr | $^{39}$Y | $^{40}$Zr | $^{41}$Nb | $^{42}$Mo | $^{43}$Tc | $^{44}$Ru | $^{45}$Rh | $^{46}$Pd | $^{47}$Ag | $^{48}$Cd | $^{49}$In | $^{50}$Sn | $^{51}$Sb | $^{52}$Te | $^{53}$I | $^{54}$Xe |
| $^{55}$Cs | $^{56}$Ba | $^{57}$La | $^{72}$Hf | $^{73}$Ta | $^{74}$W | $^{75}$Re | $^{76}$Os | $^{77}$Ir | $^{78}$Pt | $^{79}$Au | $^{80}$Hg | $^{81}$Ti | $^{82}$Pb | $^{83}$Bi | $^{84}$Po | $^{85}$At | $^{86}$Rn |
| $^{87}$Fr | $^{88}$Ra | $^{89}$Ac | $^{104}$Unq | $^{105}$Unp | $^{106}$Unh | $^{107}$Ns | $^{108}$Hs | $^{109}$Mt | | | | | | | | | |

| $^{58}$Ce | $^{59}$Pr | $^{60}$Nd | $^{61}$Pm | $^{62}$Sm | $^{63}$Eu | $^{64}$Gd | $^{65}$Tb | $^{66}$Dy | $^{67}$Ho | $^{68}$Er | $^{69}$Tm | $^{70}$Yb | $^{71}$Lu |
|---|---|---|---|---|---|---|---|---|---|---|---|---|---|
| $^{90}$Th | $^{91}$Pa | $^{92}$U | $^{93}$Np | $^{94}$Pu | $^{95}$Am | $^{96}$Cm | $^{97}$Bk | $^{98}$Cf | $^{99}$Es | $^{100}$Fm | $^{101}$Md | $^{102}$No | $^{103}$Lr |

Figure 5.1 Mendeleev's periodic table. The elements in the first two rows are hydrogen, helium, lithium, beryllium, boron, carbon, nitrogen, oxygen, fluorine and neon. The atomic number is shown for each element. This represents the number of protons in the nucleus (and therefore the number of electrons in the neutral atom).

is attracted towards the proton with a force proportional to $e^2$, the square of the electric charge. Under the effect of this force, the electron tends to remain in the innermost orbit around the proton, and the atom is in its fundamental 'ground' state of lowest energy. Now, if energy is introduced by heating or illumination, the electron may move away from the proton to take up other orbits corresponding to the atom's 'excited' states. The further the electron is from the proton, the more energy it possesses. If the electron finally acquires sufficient energy – the energy of ionisation – it jumps free of the attraction of the proton. These two parameters – the fundamental orbit and the energy of ionisation – depend directly on the fine-structure constant $\alpha$.

Physicists usually represent the successive electronic states of the atom by a scale showing irregularly spaced lines, with energy values increasing from bottom to top (Figure 5.2a). In this representation, the excited electron loses energy as it descends from one level to another. The energy lost by the electron is carried away by a photon. According to the Planck-Einstein relation, that we met in Chapter 3, the frequency of this photon is proportional to the difference in energy between the two levels in question. Photons of the same frequency make up a spectral line. Each line is characteristic of an electronic transition, and the set of lines, or spectrum, represents the succession of energy levels within the atom.

Three renowned physicists studied the lines of the hydrogen atom in the late nineteenth and early twentieth centuries: Theodore Lyman, Johann Jakob Balmer and Louis Paschen, after whom three series of lines are named. The Lyman series lies in the ultraviolet, and corresponds to transitions from excited levels down to the fundamental level. The Balmer series lies mostly in the visible domain, and corresponds to transitions down to the first excited level (Figure 5.2b). The Paschen series, in the infrared, corresponds to transitions down to the second excited level. Bohr successfully explained these series of spectral lines, so let us see how $\alpha$ can influence the positions of these lines.

## FINE STRUCTURE...

Quantum mechanics has considerably revised and improved upon Bohr's model, to the extent that we can now explain the structure of the energy levels in the hydrogen atom in very great detail. In fact, the Bohr model is too simple, because it does not incorporate relativistic effects resulting from the speed of the electron in relation to the proton: a justifiable point when we compare the speed of the electron to the speed of light. We take the speed of the electron to be the Bohr velocity $v_B = e^2/4\pi\varepsilon_0 h$, and the ratio of the two speeds $v_B/c$ as $2\pi/\alpha$, which is approximately 6/137. Physicists normally consider speeds a hundred times less than that of light to be non-relativistic.

However, the remarkable precision of atomic spectroscopy reveals a structure of levels which differs slightly from the structure calculated in the non-relativistic approximation. Certain energy levels are displaced by a small

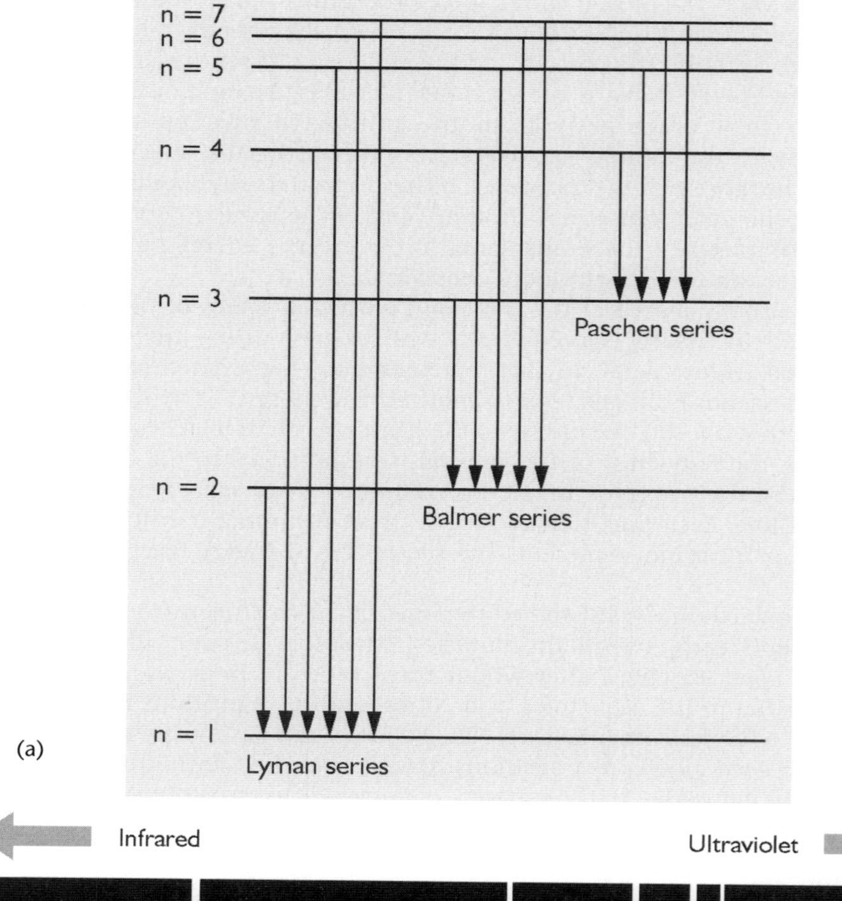


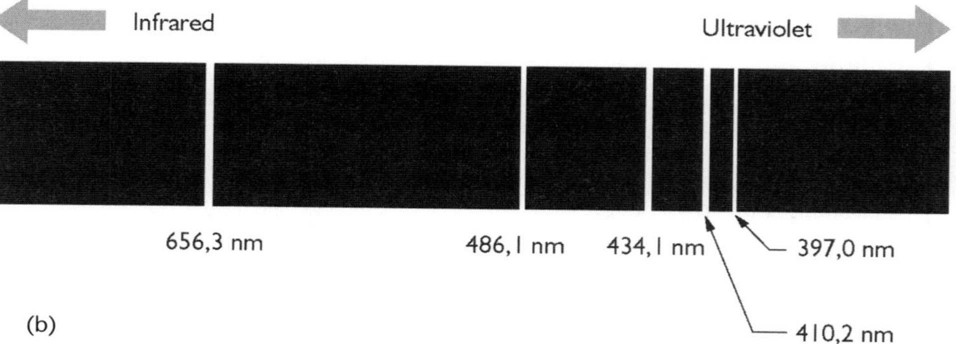


Figure 5.2 (a) Energy levels of an electron in a hydrogen atom, according to the Bohr model. The numbers represent successive orbits. The arrows indicate the transitions of the electron from one state to another. Each transition is associated with the emission of a photon, of energy equal to the difference between the two levels; to each of these corresponds to a spectroscopic line with a specific wavelength (or frequency). (b) As an example, the Balmer series of lines of the hydrogen atom (in nanometres).

amount, and others are split into various sub-levels; this has been called the 'fine structure' of an atomic spectrum, and can only be explained if we take relativistic effects into account.

Among other relativistic effects, we must consider the coupling between the 'spin' of an electron and its orbital moment. We can better understand this by borrowing images from classical physics, although that may no longer apply here. In our somewhat inappropriate imagery, we consider the electron as a charged particle rotating in the manner of a spinning top. This is its 'spin', or intrinsic angular momentum which we will represent as a little magnetised needle. The electron moves within the electric field of the proton which affects its spin, as if it were a compass needle. This interaction between the spin of the electron and the electromagnetic field of the proton, although much weaker than the electrostatic interaction, is responsible for the fine structure (Figure 5.3).

The gaps between the levels of the fine structure depend directly upon the fine-structure constant (hence the name!). Any variation in the value of $\alpha$ would reveal itself by the displacement of the levels of the fine structure of atoms.

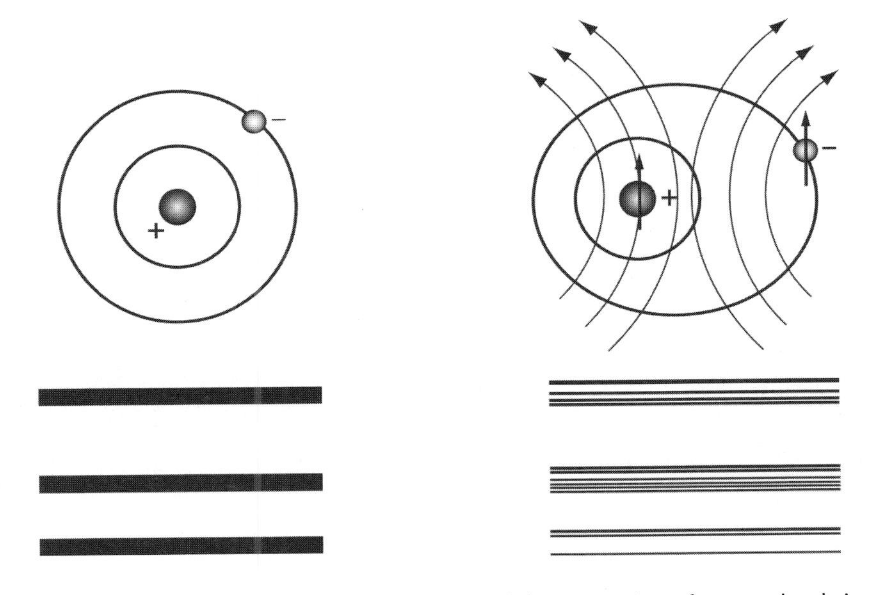


Figure 5.3 Bohr's model gave a broad picture of the succession of energy levels in a hydrogen atom (left). In reality, the hydrogen atom exhibits a 'fine structure' (right), which can be explained theoretically if one takes into account the interaction between the spin of the electron and the electromagnetic field through which it moves.

## . . .AND YET FINER

The electron is not the only particle to have spin. The proton also has spin, which interacts with that of the electron. This coupling gives rise to a further increase in the number of spectral lines of hydrogen. To improve our understanding of this interaction, let us return to our simplistic image of the magnetised needle. In the hydrogen atom, the interaction of the 'magnetised needles' of the electron and the proton causes them to align. The orientation of this alignment can occur in two different ways: either both 'needles' point in the same direction (parallel), or in opposite directions (antiparallel). In the first case, the two needles will repel each other, while in the second they will attract (Figure 5.4). The electron will therefore be closer to the nucleus if its spin is antiparallel to that of the proton, and further away if its spin is parallel to it.

The coupling between the spins of the electron and the proton is infinitesimally small. It is even smaller than the coupling involved in the fine structure. Even though the term 'coupling' seems hardly applicable, it does induce the splitting of the fundamental level of the hydrogen atom into two sub-

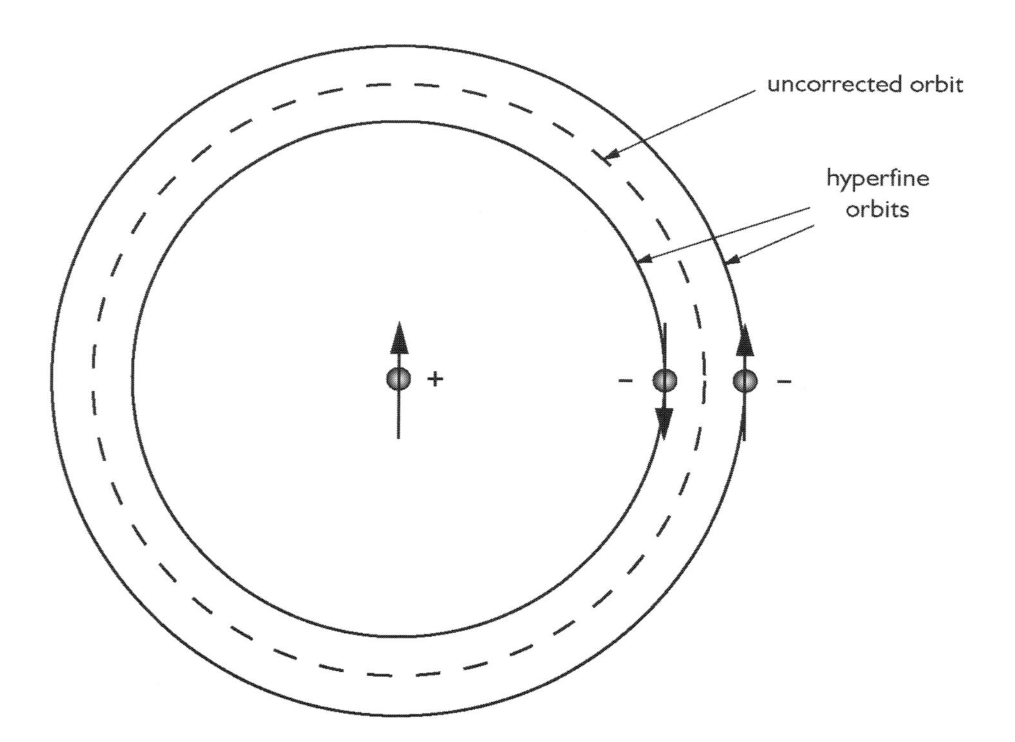


Figure 5.4 The interaction between the spins of the electron and the proton is responsible for the hyperfine structure of the fundamental level of the hydrogen atom. This level is divided into two sub-levels.

levels, known as hyperfine, one for each relative orientation of the spins of the electron and the proton. The distance between these hyperfine levels also depends upon the fine-structure constant. The transition between these hyperfine levels is detectable in the radio domain, at a frequency of 1420 megahertz, that is, at a wavelength of 21 centimetres. It is this easily observable 21-centimetre spectral line that astronomers use to detect the presence of hydrogen in interstellar space.

The hyperfine transition is used not only by radio astronomers; it has proved particularly useful in the measurement of time, since it vibrates a billion times a second. Nevertheless, physicists have not chosen hydrogen as the medium for their atomic clocks. Let us see why.

## ATOMIC CLOCKS

Atoms other than hydrogen have more complicated atomic structures, with more than one electron. Therefore, couplings with the proton are numerous and difficult to calculate. However, one family of chemical elements shows a structure similar to that of hydrogen: they are the alkali metals, which are, in order of increasing mass, lithium, sodium, potassium, rubidium, caesium and francium. In these atoms, all the electrons are grouped near the nucleus, except for one, which, unlike the others, orbits further out (Figure 5.5). The internal electrons 'couple' in pairs with antiparallel spins, such that their total magnetic effect remains negligible. Only the spin of the external electron, known as an 'unpaired' electron, interacts with that of the nucleus. Here, then, we find a configuration similar to that of the hydrogen atom, creating a hyperfine structure in the ground state.

Of all the alkali metals, caesium's hyperfine transition produces photons of the highest frequency: more than 9000 megahertz. It is therefore an excellent candidate to define the standard second, announced in 1967: "The duration of 9 192 631 770 periods of radiation corresponding to the transition between the two hyperfine levels of the ground state of the caesium-133 atom." Caesium has other advantages. It is 'pure', in that its nucleus always has the same number of protons and neutrons, unlike hydrogen, which also exists in other, heavier forms (isotopes): such as, deuterium, with a nucleus comprising one proton and one neutron, and tritium, whose nucleus contains one proton and two neutrons. The hyperfine frequency of an isotope, being slightly different, would not provide the required accuracy. Caesium is a soft metal which is liquid at temperatures above 28°C.

The idea of using caesium in an atomic clock was first proposed by the physicist Isidor Rabi in 1945 at a meeting of the American Physical Society. Rabi had just received the Nobel Prize for physics in recognition of his work on the magnetic properties of atomic nuclei. His idea was materialised 10 years later, in the National Physical Laboratory in England.

How do we make an atomic clock? Firstly, we heat caesium until it vaporises,

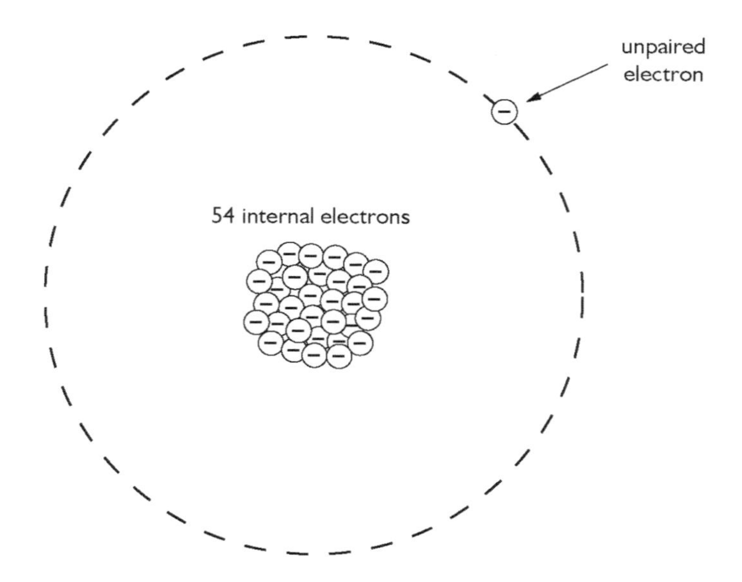


Figure 5.5 Schematic structure of the caesium atom. It has one unpaired electron, and all the others cluster around the nucleus. Its configuration is reminiscent of the hydrogen atom.

and produce a 'jet' of caesium atoms. These atoms pass through a tuneable microwave cavity, causing them to jump from one hyperfine state to another. By adjusting the frequency of the microwave radiation, we can maximise the number of hyperfine transitions, and this optimal number (or 'resonance') indicates the hyperfine frequency. The high sensitivity with which the resonance can be adjusted guarantees a great accuracy in the measurement of the frequency of the resonance. As this frequency depends directly upon $\alpha$, let us now see how we can use this property to our advantage.

## TWO CLOCKS KEEP TABS ON EACH OTHER

The atomic clock is the ideal tool for those wishing to make a very close study of the fine-structure constant. Any variation in $\alpha$ would cause a change in frequency in the clock. But since this frequency is our time standard, how can we know if it is changing or not? The answer is simple: we bring in a second atomic clock that uses a different atom. Now, the frequency of this clock would also be modified by any change in the fine-structure constant, but in a different way. If the two atomic clocks gradually got out of step with each other, this would indicate a variation in the hyperfine frequency, which could be linked to the parameter $\alpha$. And indeed, there are physicists watching closely for any such drift!

Since 2000, several researchers across the world, including Christophe

Salomon's team at the Kastler-Brossel Laboratory at the Ecole Normale Supérieure in Paris, have been comparing frequencies of two atomic clocks – a caesium clock and one using another alkali metal, rubidium. Such an experiment is easily reproducible, and can be carried out for a period of many years. This accurate experiment has seen the ratio of the frequencies of the two clocks varying by no more than $7 \times 10^{-16}$ in four years. This ratio depends not only on the fine-structure constant, but also on other constants, such as the ratio of the masses of the proton and the electron, and the nuclear magnetic moments. If we assume that the latter are indeed constant, we may deduce that the fine-structure constant has not varied by more than one part in one million billion (or one quadrillion, $10^{-15}$) per year. Its value therefore seems very stable over a few years. In order to establish this hypothesis, researchers are comparing many clocks whose frequencies depend, in different ways, on various dimensionless constants.

A similar experiment will be conducted on board the International Space Station in 2010, as part of the European Space Agency's ACES (Atomic Clock Ensemble in Space) project. A caesium atomic clock in space will attain an accuracy that has never been possible on Earth, as it will benefit from both weightlessness and the slowing down of the caesium atoms using lasers. Its hyperfine frequency will be compared with clocks of different types. The instrument will, it is hoped, have a sufficient level of accuracy to measure a possible drift of $\alpha$ of one part in $10^{-16}$ per year.

## NUCLEAR STABILITY AND INSTABILITY

There is, however, no proof at present that $\alpha$ will not vary over periods longer than just a few years. In order to investigate this, physicists are turning their attention towards the atomic nucleus. Why the nucleus? Because its stability is ensured by a positive balance between opposing forces: the electromagnetic force on the one hand, and nuclear forces on the other. Since the strength of the electromagnetic force is dictated by the fine-structure constant, nuclear reactions will provide a method of testing it.

Nearly all the mass of an atom is concentrated into a minuscule nucleus 10 000 times smaller than the atom itself. In other words, the nucleus of the atom is made of matter that is so extremely dense that a thimbleful of it would weigh more than 100 million tonnes. The nucleus consists of protons and neutrons, known collectively as nucleons. The mass of a neutron is comparable to that of a proton, but the neutron has no electric charge. Since all the protons are positively charged, the electromagnetic force keeps them separated. On the other hand, the neutrons and protons experience a different type of attraction – a strong nuclear force. This strong force is the third fundamental interaction that we encounter after gravitational and electromagnetic forces. As its name indicates, attraction between nucleons is strong, but only at short distances, less than the dimension of the nucleus. Beyond this distance the electromagnetic force prevails.

For a nucleus to be stable, there must be equilibrium between the protons and the neutrons. The more protons there are in the nucleus, the more neutrons there must be. However, an element may exist in various forms, known as *isotopes*. We have already met the two isotopes of hydrogen: deuterium and tritium. All isotopes of an element possess the same number of protons and electrons, but have different numbers of neutrons. Isotopes are characterised by the total number of nucleons they possess – their 'mass number', which is 1 for hydrogen, 2 for deuterium and 3 for tritium. Uranium, a heavy nucleus with 92 protons, has many isotopes, seven of which are long-lived. The mass numbers of these seven isotopes are 230, 232, 233, 234, 235, 236 and 238. Normally, the mass number appears after the name of the element in question, for example, 'uranium-235'.

Within the precise construction that is Nature, when a nucleus is encumbered with too many nucleons, it decays, and other, more stable nuclei are formed. An unstable isotope will decay in three different ways. 'Alpha decay' (the name has nothing to do with the fine-structure constant) involves the emission of a helium nucleus, meaning that the original nucleus has 'unloaded' two protons and two neutrons. Polonium-210, for example, decays to become lead-206. The Greek alphabet also supplies the next term, 'beta decay', which involves the transformation of one of the protons of the nucleus into a neutron or *vice versa*. This is the case with bismuth-214, which becomes polonium-214. More rarely, an unstable nucleus may split into a number of smaller nuclei: this is the third type of decay, known as fission. For example, uranium-238 will break down naturally into molybdenum-102 and tin-134. In these three types of decay (alpha, beta and fission), electrons are rearranged throughout the atom, leading to the emission of energetic photons: gamma-rays.

## CHAIN REACTIONS

Although such decays emit considerable amounts of energy, the total mass of nuclei produced is always less than the total mass of the original nuclei. The difference is equal to the binding energy of the nuclei, by virtue of the famous formula $E = mc^2$, and is released into the environment. But how can we harness and tame this energy? Those nations who were interested in the potential of nuclear energy, wasted no time in asking this question. There is, however, a problem as we cannot predict just when a nucleus will decay; the only thing we can predict is that a population of radioactive nuclei diminishes by half after a given time, characteristic of each isotope. This is known as its half-life. However, as most of the unstable nuclei in Nature have half-lives ranging from several thousand years to hundreds of millions of years, we cannot wait that long for nuclei to bestow upon us the gift of their energy! Let us see how physicists have solved this problem.

The 1930s were the most productive years for discoveries concerning the atomic nucleus. In 1932, British physicist James Chadwick discovered the neutron. In 1933, Italian physicist Enrico Fermi proposed the theory of beta decay, having realised that it involved a fourth fundamental force: the weak nuclear force. In 1934, French scientists Irène and Frédéric Joliot-Curie created new unstable elements, which decayed rapidly through beta decay. In 1938, in Germany, Otto Hahn and Fritz Strassmann discovered that the absorption of a neutron by uranium-235 led to its fission and the release of a considerable amount of energy. In 1939, Joliot demonstrated that this fission was accompanied by the emission of two or three neutrons, which could in their turn cause other fission events... Eureka! A (controlled!) chain reaction seemed possible, and Joliot went on to create, with colleagues, an apparatus containing uranium-235 together with a moderator. A moderator is a substance (water or graphite) capable of slowing down emitted neutrons in order to facilitate their absorption by other uranium-235 nuclei. This principle is used in our nuclear reactors, within which a continuous series of decays produces energy that can be harnessed.

After the Second World War, the French government set up the Commissariat à l'Energie Atomique (CEA). Its brief was to bring to the public the benefits of this formidable nuclear energy. In order to do this, it was necessary to mine uranium. So, in 1956, France began to work various uranium mines, especially in one of its West African colonies, Gabon (Figure 5.6). Our commission of inquiry will now move to Oklo, to one of the uranium deposits in the Franceville region, about 440 kilometres from the Atlantic coast of Gabon, and there we will continue our quest for the fine-structure constant. On with the tropical kit!

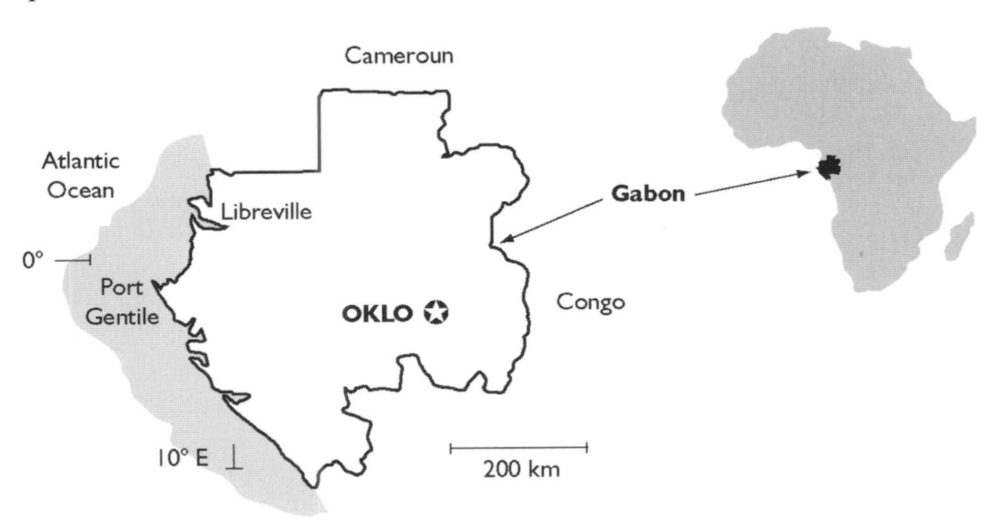


Figure 5.6 Gabon, showing the location of the Oklo mine.

## SPY STORY

The CEA was extremely interested in the very rich uranium ore deposits at Oklo. Normally, uranium from such deposits is mixed with other elements in amounts of between 0.1 and 0.5 per cent of the total; at Oklo, the proportion is between 10 and 40 per cent. After the ore has been extracted, it has to be concentrated by chemical processes in order to be suitable as a fuel. The resulting substance is a uranium oxide paste, known from its colour as 'yellowcake'. Yellowcake consists mostly of uranium-238, with a tiny admixture (0.711 per cent) of uranium-235. This latter isotope is used in chain reactions, which makes it of great interest to industry. As reactors will function on fuel containing between 3 and 20 per cent uranium-235, it is necessary to enrich the yellowcake in this active isotope if it is to be used in a reactor. The enrichment of uranium is an important (and costly) step in the chain of production of nuclear energy.

In June 1972, the CEA was obliged to check on the isotopic composition of its yellowcake: a customer of the mining company had complained that less than the expected 0.711 per cent uranium-235 was being found in deliveries. The investigation proved that the amount of uranium-235 was indeed down. In samples taken from the mine, uranium-235 was present in amounts as low as 0.4 per cent. Where had it gone?

The CEA did not release this news at once, since, in the Cold War of that time, mistrust was the rule. The inquiry was held on site. Had Oklo's uranium already been used as fuel in a nuclear reactor? There was no sign of any suspicious extraction. There was a theory that there could have been a nuclear installation at the site, which had caused contamination, but no trace of that was found. Other, more far-fetched ideas emerged: had enemy nations carried out atomic tests in the mine? Had antimatter from space fallen at Oklo? Some even wondered if extraterrestrials had landed there.

Finally, one theory, more or less ruled out at first, prevailed: the Oklo deposit had consumed some of its own uranium-235, acting as a natural nuclear reactor. This hypothesis was based on a theoretical prediction made in 1956 by Japanese physicist Paul Kuroda. Meticulous measurements of the isotopic compositions of various radioactive elements led to the reconstruction of an astonishing scenario.

## A NATURAL REACTOR

A natural nuclear reactor? Can it be possible? Yes, if an extraordinary set of physical circumstances exists. Let us look back at the history of the Oklo uranium mine.

Like all the other elements present on Earth, the uranium was forged inside stars, then dispersed into interstellar space, later to be incorporated into the material of which the solar system is composed. When the Earth was formed, about 4.5 billion years ago, the proportion of uranium-235 to uranium-238 in natural uranium was about 17 per cent. Since both these isotopes are radioactive,

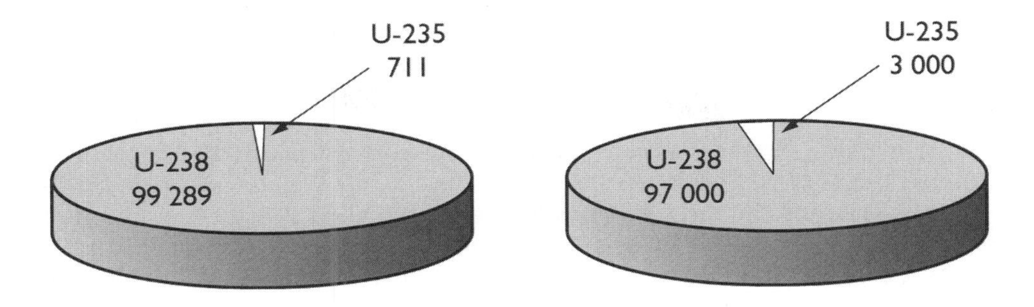


Figure 5.7 Mean isotopic composition of uranium in the solar system, showing numbers of atoms per 100 000, today (left) and 2 billion years ago (right).

their population decreases with time, but at different rates: with a half-life of 704 million years, uranium-235 disappears more rapidly than its isotope uranium-238, whose half-life is 4.47 billion years. This is why uranium-235 is no longer present today in quantities of more than 0.711 per cent (Figure 5.7).

However, two billion years ago, natural uranium still contained more than 3 per cent of uranium-235, an amount that could support chain reactions if the deposit were sufficiently concentrated and dense. That is precisely what had happened at Oklo: rainwater had been instrumental in concentrating large volumes of uranium. During this period of the Earth's history, the first unicellular life capable of photosynthesis had already developed and was releasing oxygen into the atmosphere. Carrying oxygen with it, rainwater filtered down from the surface into the ore, oxidising the uranium and making it soluble and more easily transported. Dissolved and drawn down by the water, the uranium oxide sank until it reached an impermeable layer, where it was deposited. It then formed a very large pocket of concentrated uranium ore containing enough of the isotope uranium-235 to trigger chain reactions.

This self-sustaining reactor required one other necessary condition – a neutron moderator – which was provided by the water as it continued to seep down in sufficient quantities. Slowed down by this moderator, the first neutrons emitted were confined within the uranium pocket, where they caused further fission to occur.

It seems that this remarkable combination of particular conditions happened at several sites where deposits lay. Each 'reactor' produced its slow emissions for about a hundred thousand years, in the form of heat, its output being approximately 10 kilowatts (100 000 times less than that of a nuclear power station).

## A CONSTANT CONSTANT!

Our journey to Gabon was undertaken as part of our inquiry into the fine-structure constant. What can the Oklo reactor tell us about this constant? A young Russian physicist, Aleksandr Shlyakhter, provided the answer to this question in 1976. He was a nuclear security specialist, with particular expertise in nuclear accidents such as the one at Chernobyl. Shlyakhter realised that Oklo could tell us a lot about nuclear reactions that occurred two billion years ago, provided that we knew how to interpret the concentrations of the various isotopes present in the deposit.

Let us examine the most significant case. In the mine, the concentration of samarium-149 was less than would be expected. The isotope samarium-149, a fission product of uranium-235, had later been modified, following its exposure to slow neutrons in the natural reactor. In the very specific conditions of the Oklo mine at that time, samarium-149 would capture a neutron and produce the isotope samarium-150. This is why a considerable quantity of samarium-149 had disappeared from the mine; it is in fact 45 times less abundant at Oklo than in other mines. As a result, this abnormally low concentration tells us about the probability of neutron capture at the time when the natural reactor was running.

Shlyakhter demonstrated that a samarium-149 nucleus captures a neutron more readily if the neutron's energy attains a precise value. Here we have a resonance phenomenon similar to that found in atomic clocks. The formation of samarium-150 is facilitated if the sum of the energies of the neutron and of the samarium-149 is exactly equal to the energy of a certain excited level of samarium-150. A small variation in the resonance energy would have spectacular consequences: the position of the excited level of the samarium-150 is very sensitive to the value of $\alpha$, and the same is true of the probability of the capture of a neutron by the samarium-149 nucleus. Given these facts, it was only necessary to compare *in situ* measurements and the theoretical expression of this probability to obtain a value for $\alpha$ two billion years ago.

The solution to this complex jigsaw puzzle – particularly by Thibault Damour and Freeman Dyson in 1996, with pieces from the fields of nuclear physics and geochemistry – showed that this value has not varied by more than 1 part in 10 million ($10^{-7}$) over the last two billion years. Let us note that.

## EVIDENCE FROM METEORITES

A constant value for two billion years? That's fine. But could it have varied over a longer period? Nuclear physics offers other clues. As we have seen, the stability of atomic nuclei results from the equilibrium between electromagnetic repulsion and nuclear interactions. Consequently, if the fine-structure constant had varied in the past, the decrease in radioactivity of unstable nuclei would have obeyed different laws from those observed today. If this were indeed the case, we would be able to seek verification in the relative concentrations of rhenium-187 and osmium-187. As James Peebles and Robert Dicke emphasised in 1962, this pairing was without doubt the most sensitive to any possible past variation in the fine-structure constant.

In theory, the beta-decay process transforms rhenium-187 into osmium-187 with a very long half-life of a hundred billion years, nearly ten times the age of the universe – an eternity! Let us assume, however, that the fine-structure constant had a smaller value in the past. In our hypothesis, the electromagnetic force would have been less intense and the half-life of rhenium-187 would have been even longer, making it a stable isotope. If, on the other hand, the value of the fine-structure constant had been greater, the rhenium-187 would have decayed more rapidly, with a shorter half-life. If either case had occurred, we could verify it by measuring the abundance ratios of these two elements in ancient rocks.

Such measurements have been carried out not only with terrestrial rocks, but also with extraterrestrial ones: in this case, iron meteorites. The results showed that the half-life of rhenium-187 has remained unchanged since the formation of the solar system, about 4.5 billion years ago, which means that the fine-structure constant has not varied over that period by as much as 1 part in 10 million ($10^{-7}$). We must, however, approach this measurement with certain reservations, since the age of the iron meteorites has not been determined directly. Models of the formation of our solar system indicate that these meteorites formed at the same time, to within a few million years, as meteorites of the type known as angrites. The age of angrites can be obtained through radioactive dating, thanks to the presence of the iron-207/iron-206 pair, who have a half-life that is not very sensitive to the value of $\alpha$. All these measurements seem to point to the fact that the fine-structure constant does not change with time.

Nevertheless, our commission of inquiry (and that of the physicists) is not so easily convinced! Just think about it. The last conclusion is based on too many theoretical hypotheses for our liking: the semi-empirical model of the nucleus, the model of the formation of the solar system, etc. More importantly, if we accept that the fine-structure constant has not varied since the birth of our planetary system, this does not prove that it has always been constant for an indefinite time in the past.

In order to complete its historical overview of the fine-structure constant, our commission needs to turn towards the cosmos itself. With their gaze fixed upon the stars, physicists are pulling everything out of the bag to search for the secrets of $\alpha$. The next chapter will recount this adventure.

# 6

# A constant in the clouds

Our commission of inquiry dreams of travelling back in time, to see if the fine-structure constant has varied in the past. To this day, the only time machine we have is the observation of the cosmos, and we can look deeply into space to reveal the universe as it was billions of years ago. We shall now turn our attention to outer space, so bring out the telescopes!

We are going to extend our study of the fine-structure constant through a spectroscopic observation of stars and galaxies. To this end, we have to hunt for those objects that are best suited to this: distant systems, and whose atoms emit easily identifiable spectral lines of wavelengths that are critically dependent on the fine-structure constant.

Before we make the acquaintance of the objects that are most suited to our purpose, we need to draw up an inventory of all that is up there, above our heads. The further out we have pushed the boundaries of our knowledge of the universe, the more we are reminded of the small scale and relative insignificance of our own being. Our Sun is just another star among the tens of billions that populate the Milky Way among the teeming multitudes that amazed Galileo at his telescope. Even traversing the Milky Way, the distances involved would create a sense of vertigo. Within Kant's "world of worlds" (as he called our Galaxy), the nearest star to the Sun, Proxima Centauri, is more than four light-years away. Conversations with any inhabitants in its vicinity would not be easy, as each exchange of information would take over eight years. Light takes 30 000 years to reach us from the centre of our Galaxy, and today we are observing it as it was when our ancestors were creating Europe's oldest cave paintings in the Grotte Chauvet.

However, the universe extends far beyond the Milky Way. Other galaxies are whirling through the cosmos, and the nearest – one of the Magellanic Clouds – is about 160 000 light-years removed from Earth. Since the early decades of the twentieth century, astronomers have discovered billions of galaxies, separated by millions of light-years. Across this vastness of space and time, galaxies occur in clusters a few thousand strong. The nearest of these is the Virgo Cluster, about 60 million light-years distant. Its light dates back to the time of the emergence on Earth of the first primates, so let us explore the past by the light from the stars.

## SURVEYORS OF THE SKY

How do astronomers determine such distances? When we look up at the night sky, we see luminous points distributed across its vault, as in a planetarium, yet they all appear to be at similar distances. In order to measure the distances of the planets, and of nearby stars, we resort to methods involving triangulation or parallax: we observe the same heavenly body from two different places, and apply trigonometry to deduce its distance. However, this method works only at a limited range, and cannot be applied to the measurement of the more distant universe as the angles become too small to be measured.

Different scales of distance require different methods. To calculate the distances of galaxies, we look within them for the presence of certain variable stars known as *Cepheids*. The brightness of a Cepheid rises and falls regularly, with a period dependent upon its absolute luminosity. Since the apparent luminosity of a star decreases with its distance, its period of variation gives an indirect clue to that distance. In 1912, American researcher Henrietta Leavitt proposed this method of measuring the distances to galaxies, and her work led to the production of the first catalogues of galaxies.

Using these catalogues in the 1920s, the great American astronomer Edwin Hubble noticed an odd phenomenon. He observed that the lines in the spectra of galaxies were systematically shifted in comparison with laboratory spectra. Their wavelengths all seemed longer than expected. Interpreted as a result of the Doppler–Fizeau effect, this *redshift* indicates that the galaxies are moving away from us. We are all familiar with the Doppler effect in the realm of sound: if a car or a train passes us, we hear a higher note (frequency) when it is approaching us and a lower note when it is moving away. Similarly, if light from a luminous source appears redder, it is because that source is moving away from us. Therefore, the redshift in spectra indicates that the galaxies that are emitting the light are receding. Hubble also determined that the more distant the galaxy, the greater will be the observed redshift. He had therefore established a law of proportionality governing the distance of galaxies and their speed of recession. Hubble had introduced a new constant, the Hubble constant, which has the dimension of the inverse of time, in order to relate the recession velocity and the distance of the galaxy.

This general motion suggests an ever-changing, expanding universe. In fact, such redshifts had already been predicted by Belgian theoretician Georges Lemaître, in an article published in 1927. He put forward an original solution to the equation of general relativity, implying a dynamic universe that is born in a condensed, hot state, which cools as its volume is increasing. As a consequence of this universal expansion, the atomic spectra of distant objects would exhibit a redshift, or rather a dilatation: as light propagated towards our instruments, all wavelengths would increase, in the same manner as all spatial distances (Figure 6.1). By sheer habit, we continue to use the term 'shift' even though it is misleading.

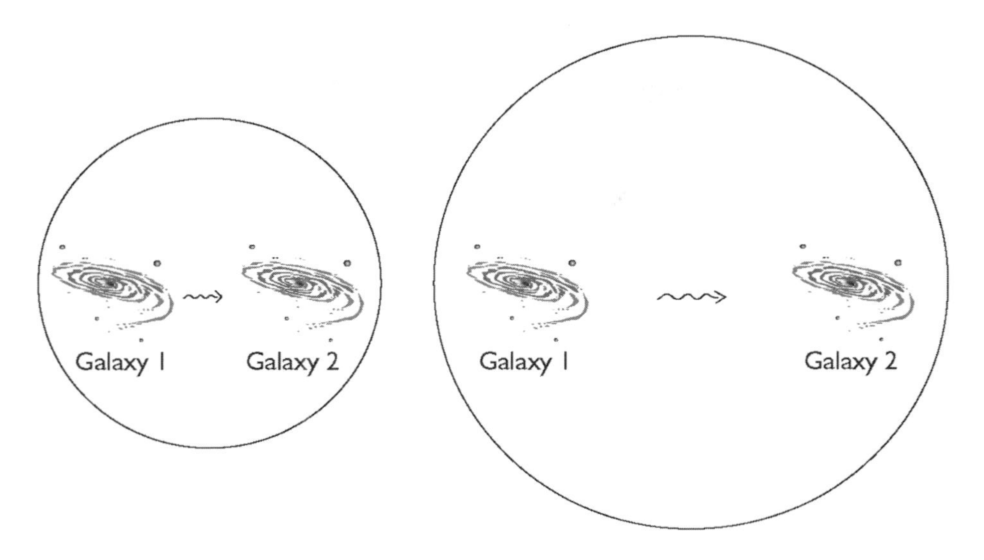


Figure 6.1 The redshift in the light emitted by distant galaxies is due to the expansion of the universe.

Lemaître's prediction went unnoticed, for various reasons. Firstly, his article was in French, whereas the scientific community normally corresponded in English. Then, when British scientist Arthur Eddington translated the article in 1931, he omitted the prediction of the redshift. Finally, most physicists, accustomed to dealing with a static universe, were loath to accept a dynamic model. However, British scientist Fred Hoyle disdainfully coined the term 'Big Bang' for it – a label that caught on and finally became the accepted phrase when referring to the 'standard cosmological model'. Eventually, this model prevailed, and was independently refined by physicists such as Gamow, Hermann and Alpher.

Since Hubble and the advent of the Big Bang model, astronomers have used the redshift method to calculate the distances of galaxies: some of the order of millions or billions of light-years. The redshift, written as $z$, is zero if the observed spectrum is the same as the laboratory spectrum. Its value is 0.01 if the wavelength of the lines is shifted by 1 per cent when compared with the laboratory spectrum: this translates into a distance of about 1 billion light-years. If $z = 0.1$, the wavelength is shifted by 10 per cent, indicating a distance of 2 billion light-years. If $z = 1$, the wavelength is doubled (100 per cent), and the distance is 5 billion light-years. If $z = 2$, the wavelength is tripled and the distance becomes 10 billion light-years, and so on. The redshifts of the most ancient galaxies have been found to be in the order of $z = 10$.

## SPECTRAL SHADOWS

Does our inventory of the cosmos include only galaxies? Astronomers long believed that the universe was a huge, almost empty space, containing scattered stars and galaxies separated by nothingness. That is far from reality. As a result of developments in spectroscopy and radio astronomy, dark lines were discovered in stellar spectra in the 1920s. These lines revealed the presence of a dusty interstellar medium, albeit billions upon billions of times more tenuous than the dust in our homes. This medium consists of atoms, molecules, grains and particles of dust at various densities and temperatures, all of which could come together at some future time to form new stars. Interstellar dust clouds are too dark for us to observe directly, but when they lie between us and the starry background, they intercept certain frequencies of light and cause dark lines to appear on stellar spectra: shades of Corneille's "obscure brightness" (see Introduction). How does this occur?

Some of the photons emitted by a star will be of exactly the right wavelength to raise certain electrons to a higher energy state within their atoms. All that is required is for the wavelength to be equal to the difference in energy between the two relevant levels, divided by Planck's constant (according to the Planck–Einstein relation). When an electron encounters such a photon, it absorbs it and jumps to a higher level. Shortly afterwards, it is de-excited (returning to its original level) and emits a new photon of the same frequency as the first. However, this new photon is emitted in a random direction, and there is little chance that it will move in the same direction as that taken by its predecessor, towards the telescope or spectrometer aimed at the star. The result of this is that the light will have been subtracted, leaving a dark line in the spectrum of the star in question, at the characteristic wavelength of the electronic transition.

These dark lines constitute an absorption spectrum – a veritable negative of the emission spectrum encountered in the previous chapter. Both spectra carry similar information; the absorption spectrum shows the presence of atoms or molecules between the star and the observer, and the frequencies or wavelengths of the lines reveal the identity of these atoms and molecules. Absorption spectra of interstellar clouds have provided evidence of large quantities of atomic hydrogen within the Milky Way. Characteristic of this hydrogen is a hyperfine line at a wavelength of 21 centimetres. In the interstellar medium, astronomers have discovered not only simple molecules but also sodium and heavier elements such as calcium.

Not long after their discovery of interstellar clouds, astronomers also realised that the intergalactic medium was not quite as empty as they previously thought. It holds about ten atoms in every cubic metre, which may not seem to be much but is enough to play a part in the formation of galaxies. While we try to comprehend the nature of these intergalactic clouds, what interests us more directly is what atoms located in remote regions can tell us about the past value of certain fundamental constants. These immense clouds of absorbing matter have provided considerable insights into the value of the fine-structure constant billions of years ago: they provide us with 'palaeospectra' – that is, spectra from ancient times.

## A BRIEF HISTORY OF PALAEOSPECTROSCOPY

To obtain the most ancient palaeospectra, we must examine the most distant sources of light. It is difficult to observe very distant luminous objects, since their apparent brightness diminishes rapidly with distance. Two effects contribute to this: the dilution in the number of photons, and the diminution of the energy of each photon due to the expansion of the universe. We must therefore search for very energetic sources. Also, as we expect their spectra to be redshifted, we must observe at long wavelengths.

In the 1950s, astronomers began to detect a large number of radio sources, which they then tried to match with known optical sources. Most of the radio sources corresponded to distant and therefore very ancient galaxies. In 1956, American physicist Malcolm Savedoff decided to use these sources to compare the values of physical constants in those remote regions with their values in the laboratory. He analysed the absorption of radiation from a radio source in the constellation of Cygnus, known as Cygnus A. This radio galaxy has a redshift of the order of $z = 0.057$, which means that it is several hundred million light-years away from us. The light emitted by Cygnus A passes through dust clouds, whose absorption lines reveal the presence of, among other things, elements such as hydrogen, nitrogen, oxygen and neon. Some of this material is ionised, but the nitrogen is doubly ionised (i.e. lacking two electrons) and the neon is triply ionised (lacking three electrons). Both nitrogen and neon show a fine structure in the radio domain.

The first stage in the analysis involves recognising lines belonging to the spectrum of the same element, and the spectrum is then compared with those obtained in the laboratory. If the first relates to the second by a simple dilatation factor (Figure 6.2), then the expansion of the universe is responsible for the observed shift of the lines. However, if other detected shifts are dependent on wavelength, we may ascribe them to a different value of the fine-structure constant at the time when the light from Cygnus A passed through the dust clouds. It therefore takes real detective work to investigate the past value of $\alpha$: first, sorting the lines into 'systems' of the same spectral shift and, second, deducing the value of $\alpha$ from the relative position of lines in the same system. It is most important to measure the *relative* displacement of lines rather than the displacement of a single line, as the latter is probably the result of the relative motion of the intergalactic cloud.

After playing detective, Savedoff concluded that the fine-structure constant had varied by no more than 1 part in 1000 ($10^{-3}$) over a period of 100 million years. Admittedly, the results from the Oklo natural reactor were more accurate and covered a longer period; but intergalactic spectroscopy would evolve, to offer even more remarkable insights.

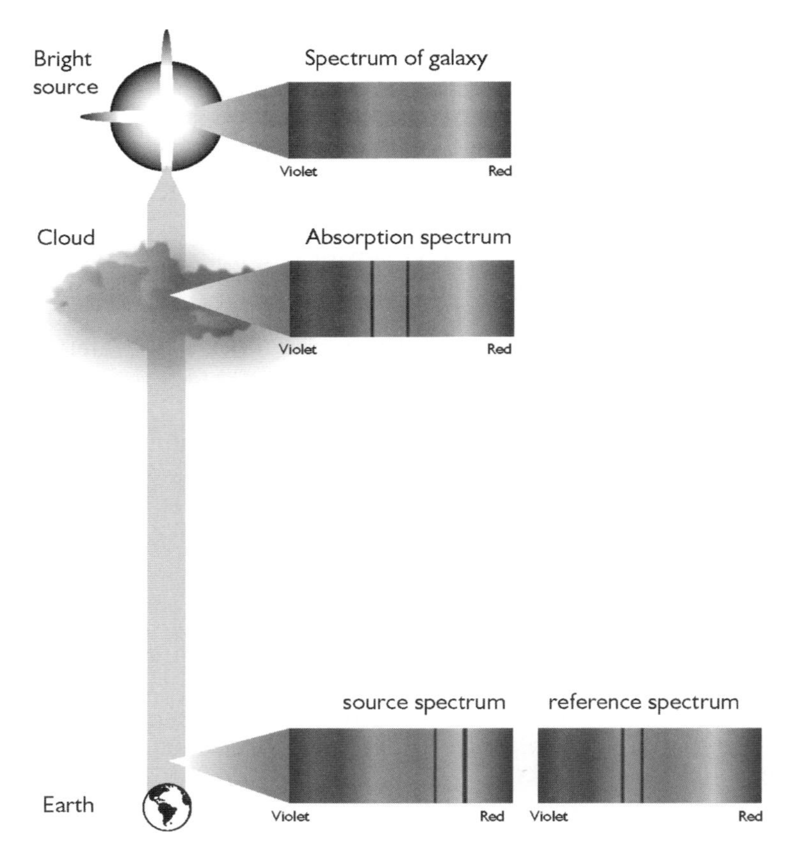


Figure 6.2 Path taken by light from a distant galaxy to an observer on Earth. The galaxy emits an almost continuous spectrum across a wide range of wavelengths. Chemical elements present in an intergalactic cloud absorb certain wavelengths, causing dark lines to appear on the spectrum. Because of the expansion of the universe, these lines are shifted towards the red end of the spectrum, as the light makes its way to Earth. The observed spectrum is compared with a laboratory reference spectrum, to ascertain whether the value of $\alpha$ has varied since the time at which light crossed the cloud. (After Uzan, *Pour la Science*, July 2002.)

## BEACONS IN THE UNIVERSE

As intergalactic spectroscopy became more refined, even older intergalactic clouds could be examined, and astronomers had discovered luminous sources beyond the radio galaxies. Among the radio sources catalogued in the late 1950s were some whose optical counterparts were not easy to see. These took the form of small, faint objects, poorly resolved, with unfamiliar spectra, and puzzled astronomers labelled them 'quasars' (quasi-stellar sources) as they were starlike, but were not stars.

These objects radiate intensely across the entire electromagnetic spectrum, from radio waves to X-rays and gamma-rays; this indicates that they are sites at which many physical processes are taking place. The efforts of astrophysicists were rewarded when they recognised, in the spectra of quasars, emission lines identifying various elements. The dominant line was due to hydrogen (the Lyman-$\alpha$ line, first in the Lyman series; see Figure 6.3a). This line was difficult to identify for two reasons: (1) it was broadened, because the hydrogen atoms are all moving relative to each other, dispersing the wavelengths of the transition, and (2) the line was considerably redshifted. The first two quasars that were found (3C 48 in 1960 and 3C 273 in 1962) showed unheard-of redshifts, of 0.37 and 0.16 respectively, indicating that these objects were further away than the most distant galaxies yet detected: 3C 48 is four billion light-years away, and 3C 273 is two billion light-years away. The cosmic time machine now travelled to epochs as long ago as the birth of the solar system, and even earlier.

Quasars therefore qualify as the furthest and most luminous objects known. Given their distances, their intrinsic luminosity must be several hundred times greater than that of the Milky Way galaxy. It seems as if a whole cluster of galaxies has been squeezed into a volume the size of the solar system. However, quasars, when observed, can be 20 000 times fainter than a star at the limit of naked-eye visibility, which is one of the reasons that make their light difficult to measure. To date, some 10 000 quasars have been catalogued; and while the edge of the observable universe is about 15 billion light-years away, the remotest of the quasars detected have been blazing away at us from distances of about 13 billion light-years. The physical nature and workings of quasars have long been subjects of debate, but the consensus among today's astrophysicists is that quasars were formed 2 or 3 billion years after the birth of the universe,

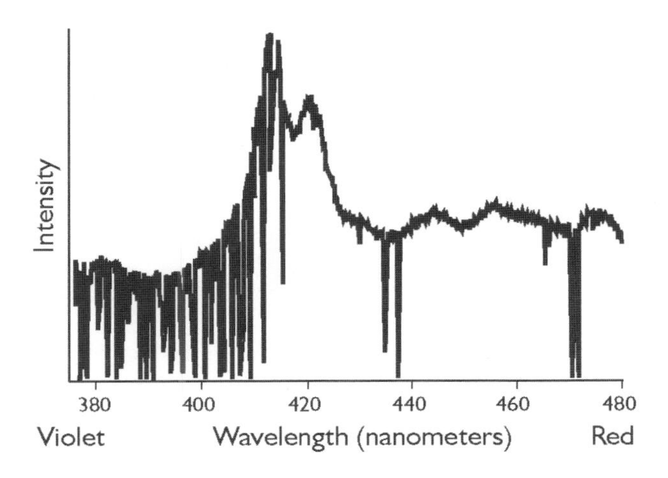


Figure 6.3 (a) Spectrum of a quasar: intensity curve as a function of wavelength.

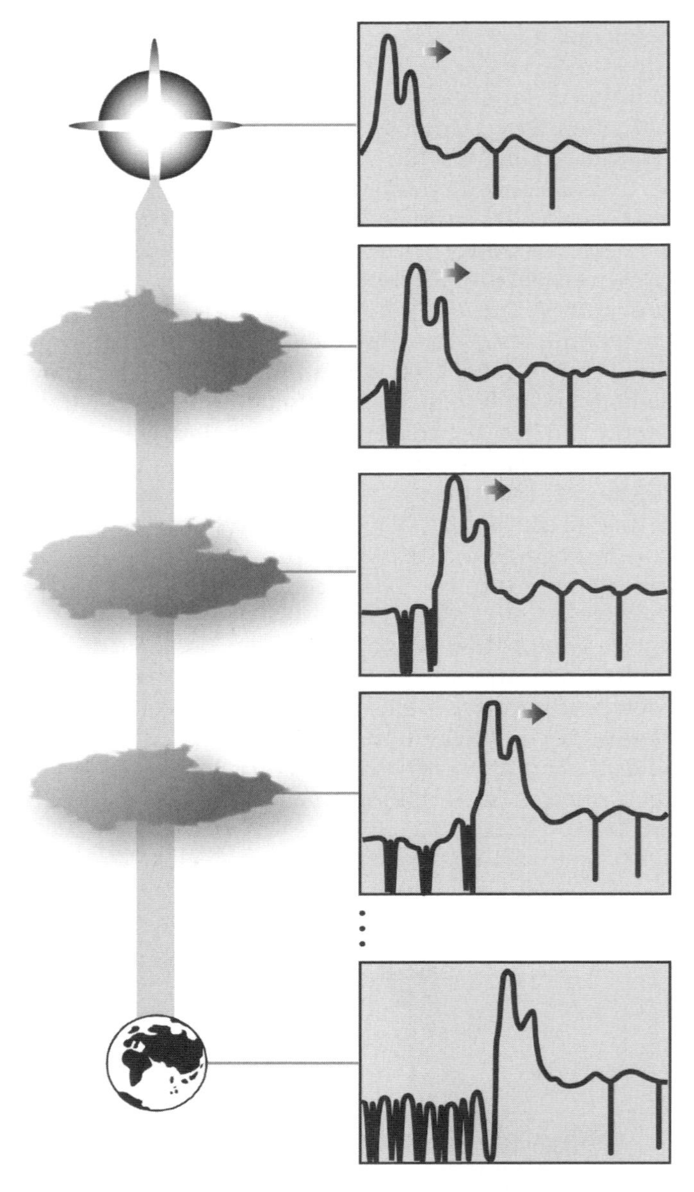

Figure 6.3 (b) The main 'spike' in the quasar's spectrum is the Lyman-$\alpha$ emission line of the hydrogen atom. It is broadened by the motion of atoms in the quasar, and redshifted as light propagates towards the Earth. Then, hydrogen in an intergalactic cloud imprints a Lyman-$\alpha$ absorption line, further redshifted, onto this spectrum. Each intergalactic cloud adds its (shifted) absorption line to the spectrum, forming the Lyman-$\alpha$ 'forest'. On the right of the spectrum we see lines due to heavier atoms (cf. Figure 6.3(a)). (After Petitjean, *Pour la Science*, November 2002.)

around supermassive black holes lodged at the centre of galaxies in the process of formation. Irresistibly drawn towards the black hole, a vast whirlpool of matter emits much of its energy as radiation. Such is the quasar.

However, it is the spectra of quasars that interest us as we pursue our inquiry into the fine-structure constant. More precisely, our focus is on the absorption lines created by intergalactic clouds between them and us, and astrophysicists have been studying these lines since 1966. Certainly, the great distances of quasars ensure that, along their line of sight, there will be very ancient intergalactic clouds; but this also means that absorption lines will be present in large numbers. Along the path of the light emitted by a quasar, each intergalactic cloud stamps its own absorption lines on the spectrum, and these lines are redshifted in proportion to the distance of the individual clouds. Among other effects, neutral hydrogen causes an absorption line at a wavelength of 121.5 nanometres (billionths of a metre) – the Lyman-$\alpha$ line – and the further away the cloud responsible for this line, the greater the shift of the line towards longer wavelengths. Astrophysicists are therefore peering into a veritable 'forest' of Lyman-$\alpha$ lines, revealing the presence of hundreds of intergalactic clouds along the line of sight of the quasar (Figure 6.3b).

In addition to these lines, there are lines due to carbon, oxygen and silicon, many times ionised. For example, triply ionised carbon ($C^{3+}$) shows two characteristic absorption lines at 154.8 and 155.0 nanometres. The detective work now involves the tiniest detail as each atom or ion has in fact many lines, of various intensities, but only some are detectable in the ultraviolet and optical domains. The component with the most lines is molecular hydrogen ($H_2$). Situated in the ultraviolet, its lines are not observable from Earth, since our atmosphere is opaque to these wavelengths. Theoretically, we therefore have to observe them using space telescopes such as the Hubble. However, if the absorbing gas is very far away, the absorption is displaced towards longer wavelengths owing to the expansion of the universe. For objects at distances of more than 5 billion light-years, the shift is so pronounced that the lines become observable in the visible domain. The expansion of the universe therefore means that we can conduct our palaeospectroscopic observations using large, ground-based telescopes.

## DOUBLETS

In 1967, three American astronomers, John Bahcall, Maarten Schmidt and Wallace Sargent, set themselves a long and painstaking task. They studied pairs of absorption lines belonging to the fine structure of quadruply ionised silicon. This ion behaves like an alkali-metal atom, its first excited level being split into two fine sub-levels. The gap between the lines of this doublet depends on the square of the fine-structure constant $\alpha$, which means that it can be used to give a precise measurement.

The three researchers investigated these doublets in the palaeospectra of clouds along the line of sight of 3C 191, a newly discovered and very distant quasar. As its observed spectral shift of 1.95 gave a distance of about 5 billion light-years, quasar 3C 191 is being seen as it was before the formation of the solar system. The study concluded that the fine-structure constant had not varied by even a few per cent since that epoch.

A group led by Dmitry Varshalovich, of the Ioffe Institute in St. Petersburg, has led the way in this field for many years. In the 1990s, they showed that variations of $\alpha$ were less than 1 part in 10 000 ($10^{-4}$). With developments in telescope technology, and new electronic detectors continually improving the accuracy of measurements, it was established by 1999 that there had been no variation in the fine-structure constant to within 1 part in 100 000 ($10^{-5}$).

Aiming at even greater precision, astrophysicists have to go beyond the existing capacity of their instruments. They have an arduous task, but what exactly are they trying to measure? They are seeking out infinitesimal displacements of lines, resulting from a small variation of $\alpha$ over time – i.e. during the journey made by the light emanating from the remote universe to our telescopes and spectrometers. If you train the eye on the spectrum of a quasar, see the Lyman-$\alpha$ 'forest', you will realise the magnitude of the task. The lines are very close together, and the spectral resolution has to be excellent. Lines have to be measurable to better than a fifth of a pixel of the detector. This means achieving an accuracy of a thousandth of a nanometre with measurements around 450 nanometres, in the very feeble light of a quasar at the edge of the universe. Even on Earth, these measurements are real technical triumphs, though not all wavelengths are explored in such a precise way. This is a limitation, since, if we wish to prove that a wavelength has varied over time, we have first to know its exact value in the laboratory. This is not always possible in, for example, the case of triply ionised carbon lines. Fortunately, triply ionised silicon lines can be measured very precisely, and the situation is also favourable when measuring lines of iron and magnesium, atoms that will play an important part in our further investigations.

## MULTIPLETS

In the late 1990s, John Webb's team took up the challenge of precise measurement and introduced a clever technique known as the 'many-multiplet' (or 'MM') method. Instead of measuring the gaps between the lines of a doublet of the same element (silicon, for example), they compare many absorption lines relating to the various chemical elements present in the clouds. They will previously have performed, on powerful computers, simulations of the displacement of the absorption lines due to small variations of $\alpha$.

They have deduced that the same variation would affect positions of different lines in different ways: one might be redshifted while another might be blueshifted, each showing a specific displacement. In particular, a variation of $\alpha$ would alter the positions of lines of quadruply ionised silicon and doubly ionised iron, but would leave magnesium lines virtually unchanged. These latter therefore serve as a useful reference system, or 'anchor point'. When one measures the shifts of other lines compared to those of magnesium, one may assume that the latter result from a variation in the fine-structure constant, and not from any systematic shifts (cancelling each other out during the comparison).

Between 1999 and 2001, Webb and his colleagues applied the MM method to a total of 28 absorption spectra from 13 different quasars, using the HIRES spectrograph at the Keck Observatory in Hawaii (Figure 6.4). They analysed positions of the Lyman-$\alpha$ lines, and those of several other ions including magnesium, aluminium and iron, in various states of ionisation. To date, their analysis, based on 128 absorption systems with redshifts ranging between 0.5 and 3, has produced data showing that the fine-structure constant has decreased in value by 5 to 10 parts in 100 000 over the last 6 to 11 billion years.

In the past, this very tiny variation would have been subsumed within the range of uncertainty in measurements. The present result takes a zero variation of $\alpha$ outside the error bars, and therein lies its novelty, and also a surprise. As long as the results were compatible with zero variation, physics was confirmed and we could trust the measurements; but since this result foreshadows a revolution in physics, we must assure ourselves that it is has a solid basis.

Figure 6.4 The Keck Telescope in Hawaii. Each telescope has a 10-metre mirror. (© NASA.)

## SPOT THE ERROR

Has the fine-structure constant really varied as the universe has evolved, as this measurement seems to indicate? As we have seen, the variation in question is tiny, and very difficult to measure. After Webb and his colleagues had published their findings, various teams of physicists and astrophysicists took it upon themselves to seek out any possible sources of error, and to evaluate the bearing they might have upon the results. The list grew ever longer... systemic errors, relating to both the physical systems and to the observational procedures, were suggested.

Among the former:

- if atoms producing the lines being compared are located in different regions of the cloud, there is a possibility that those regions are moving in different directions, which would result in various Doppler shifts of the lines, giving a spurious variation of $\alpha$;
- two isotopes of the same element present slightly different spectra, which might suggest that a gap has occurred;
- the energy levels of the atoms could have been altered by a magnetic field.

Errors suggested for the observational procedures centred around the spectrograph and its environment:

- the speed of the Earth varies during the acquisition of the data, causing Doppler effects on different spectra observed at different times;
- in the case of ground-based observations, dispersion of light within the atmosphere can cause distortion of the spectra;
- a local temperature variation can alter the refractive index of the air in the spectrograph, and affect the position of the lines.

There are obviously many possible sources of error. The best way to confirm Webb's measurement is to repeat it, if possible, using other absorption systems and ensuring that a few precautions are taken. A Franco-Indian team, under the joint direction of Patrick Petitjean (of the Paris Astrophysical Institute) and Raghunathan Srianand, set out to do this. These researchers used a state-of-the-art spectrometer called UVES, in conjunction with one of the world's largest telescopes, the VLT (Very Large Telescope). The VLT operates on one of the best observation sites on our planet, on top of Chile's high plateau (Figure 6.5). A crucial aspect of the observing programme was the imposition of rigorous criteria in the choice of the lines to record in their analysis. For example, they chose lines of doubly ionised elements, to the exclusion of all others. This meant that they would avoid comparing ions from different parts of a cloud moving relative to each other. They also eliminated absorption lines that had been contaminated by the Earth's atmosphere.

Proceeding cautiously in this way, they applied the MM method to a homogeneous sample of 50 absorption systems, along the lines of sight of 18 remote quasars. They recorded spectra for 34 nights to achieve the highest

Figure 6.5 The Very Large Telescope (VLT) in Chile. It consists of four telescopes, each with an 8-metre mirror. (© European Southern Observatory.)

possible spectral resolution and an improved signal-to-noise ratio. Their conclusion, when it came, contradicted that of Webb and his colleagues. Not only did it point to a zero variation of the fine-structure constant, but also set tighter limits upon its margins of variation: over the last 10 billion years, the variation of $\alpha$ was found to be less than 0.6 part per million ($10^{-6}$).

## TRICKS OF MAGNESIUM

So what shall we conclude from this? Which should we believe? Is there any hope of reconciling these two series of measurements or do they indeed tell us something unexpected about the early universe?

Among the possible sources of errors that we listed, there is one upon which we might profitably dwell. As we stated, two isotopes of the same element present slightly different spectra, and the observed shifts of their lines might lead us to infer a variation of the fine-structure constant. Now, both the Webb and the Petitjean results depend on hypotheses that these researchers have adopted about the isotopic composition of the element magnesium. Remember that this is the 'anchor' element for the determination of shifts in the lines of other elements. If magnesium also exhibited a shift, then that would be likely to affect all the results.

There are three isotopes of magnesium: magnesium-24 is the lightest and most abundant; magnesium-25 has one extra neutron; and magnesium-26 has two extra neutrons. Within the Milky Way, the three isotopes exist in the ratio 79:10:11 respectively. The two research teams assumed that magnesium would be present in the most distant (and ancient) intergalactic clouds, in ratios that are similar to those found in our own Galaxy. It seems a natural idea as we assume local conditions to be the same elsewhere in the universe. But is this plausible? To answer this question, we must take a close look at the formation of the chemical elements in the forges of the cosmos.

Stars create the elements (for example, carbon, nitrogen and oxygen) of which we are made, by the process of *stellar nucleosynthesis*. In the case of our Sun, hydrogen nuclei (protons) fuse together to become helium nuclei. When, as will happen in the Sun in a few billion years, the abundance of hydrogen in the core is gradually depleted, the star's internal processes will change. The core will shrink, causing the temperature to rise, and a new fusion process will be initiated, this time involving helium. Carbon, nitrogen and oxygen will be produced and the star will become a 'red giant', swollen as the energy released by the fusion reactions pushes the upper layers outwards, the red colour characterising a lower surface temperature for the star. For most low-mass stars, like the Sun, this will mark the beginning of the final stage of their lives.

Stars 10 to 15 times more massive than the Sun proceed through additional stages. After they have used up their helium, a new contraction takes place, with further fusion producing neon, sodium and magnesium. Supermassive stars, by-passing this stage, undergo different collapse mechanisms in which new elements, from oxygen to iron, are produced. Elements heavier than iron are created in the enormous stellar explosions known as supernovae, which seed the space around them with vast numbers of different nuclei.

The material emitted by supernovae and other stars floats freely through space, forming vast interstellar and intergalactic clouds. In their turn, these clouds may undergo gravitational contraction, and new stellar systems will be generated, to exist as sites of further nuclear fusion. Everything that is made inside stars is recycled in the clouds.

## THE RED GIANT REVISITED

From our study of this process, we know that elements heavier than oxygen appeared later than the lightest elements. We therefore infer that ancient regions of the universe, and especially intergalactic clouds some 10 billion light-years away from us, are poorer in metals (for example magnesium and iron) than the Milky Way. Importantly, we should realise that isotopic concentrations in these clouds are different from those in our vicinity.

From what we know about stellar nucleosynthesis, we can assume that the two heavier isotopes of magnesium – magnesium-25 and magnesium-26 – are much rarer in these ancient regions than the lighter magnesium-24. Let us disregard these heavier isotopes in distant intergalactic clouds, and focus only on the magnesium-24 found there. Our isotopic ratio for magnesium would thus be 1:0:0. If this hypothesis is applied, Webb's results become even more surprising: the variation of the fine-structure constant would be even greater than announced! Isotopic concentrations of magnesium do indeed affect a possible variation of $\alpha$, but not in the way we might have expected.

No matter ... perhaps magnesium-25 and magnesium-26 were produced in greater quantities in the past, in first-generation stars. Casting around, we come across isotopic ratios for magnesium that reconcile the measurements of Webb and Petitjean. If we assume figures of 63:37:37 for the intergalactic clouds, rather than the 79:10:11 of our Milky Way, we infer a zero variation for the fine-structure constant.

Therefore, instead of revolutionising physics and fundamental constants, the many-multiplet method might be telling us more about the way first-generation stars work. Nucleosynthesis might have been a different process in ancient stars. A new model has been proposed, in which first-generation red giants are hotter than those in our region of the universe, and thereby produce the 'right' amounts of isotopes. It will be possible to test this model on the basis of predictions arising from it, concerning isotope ratios in other elements. Watch this space...

## BEYOND QUASARS

The quasar absorption spectra open a window on the ancient universe, when it was just 2 or 3 billion years old. Can we look further back in time, to an even younger universe? Two stages in the early evolution of our universe have left imprints that are still detectable today.

In the furnace that was the universe just after the Big Bang, no structured matter could exist. As the temperature fell, the first structures began to appear. Quarks combined to form baryons such as protons and neutrons, which in turn became nuclei; then nuclei and electrons became atoms. The last two stages, known as the primordial nucleosynthesis and recombination, are accessible to investigation using modern techniques.

Again, let us mount our time machine, and travel back to the epoch of the recombination, only 300 000 years after the Big Bang. Just before this epoch, the universe had acted like a black body, with protons, neutrons and electrons constantly emitting and reabsorbing photons. Light was trapped by matter. As long as the radiation energy dominated, all atoms were ionised by the high-energy photons in which they were bathed. As the universe expanded, the wavelength, and therefore the energy, of this sea of photons decreased. Neutral atoms could now form, without being re-ionised by radiation. The universe then became transparent to photons, such that light decoupled from matter and escaped in all directions. Ever since, this radiation has flooded the whole universe and cooled as the universe expanded. It is known as the cosmic background radiation.

Gamow, Alpher and Hermann predicted in 1948 that the cosmic background radiation must, even today, echo throughout the universe at a temperature of a few kelvin: 'fossil or relic radiation', bearing witness to the fiery birth of our universe. In late 1964, two radio astronomers, Arno Penzias and Robert Wilson, discovered this cosmic background radiation by chance, and identified it as such with the help of American astrophysicists Robert Dicke and James Peebles. This discovery represents one of the most tangible proofs of the Big Bang model.

The cosmic background radiation has the profile of black-body radiation at a current temperature of 2.73 kelvin (–270 °C). It is almost uniform and isotropic, i.e. the same in all directions in space. Knowing that electromagnetic interactions prevailed during the recombination, we can assess the value of the fine-structure constant from radiation emitted at that time. Was its value the same as today's? To find the answer to this question, astrophysicists have been examining the cosmic background radiation in very great detail, thanks largely to observations by two satellites: COBE, with its excellent radiometers and, more recently, WMAP. They have managed to measure slight irregularities and anisotropies in the radiation, which bear witness to the origins of the large-scale structures which formed. These measurements suggest that the fine-structure constant has varied by no more than a few parts in a thousand since the epoch of the recombination. It has to be said that the accuracy of these measurements is not of the best, and what is more, the method depends on other cosmological parameters such as Hubble's constant and the proportion of protons and neutrons in the universe, which must be determined independently. Perhaps the accuracy will be improved in the future.

There are faint traces of an even younger universe: the present abundance of light nuclei such as hydrogen, deuterium, helium and lithium offer some insight into the primordial nucleosynthesis. These abundances provide another clue in favour of the Big Bang hypothesis. Between one second and a few minutes after the primordial explosion, protons and neutrons combined to form the lightest elements. The abundances of these elements depend on a number of parameters, one of which is the fine-structure constant. By applying various hypotheses to these other parameters, several researchers have obtained a result similar to that of the work done on the cosmic background radiation. However, physicists are not completely certain of this result, and the picture remains unclear.

## VARIATION ON ANOTHER THEME

Let us take stock. For most of the time that the universe has existed, the fine-structure constant does not seem to have varied. The only measurement ascribing a slight variation to it has been that of Webb and his colleagues, but this has not been confirmed by later measurements. Moreover, this measurement could be explained away if we assume that stellar nucleosynthesis has not always been the same in the past as it is today. Further tests of the fine-structure constant are needed, and researchers are applying themselves to the task.

There is, however, another dimensionless parameter that is arousing interest: Petitjean's team has detected signs of a variation in the parameter $\mu$, the ratio of the masses of the proton and the electron. Petitjean and his co-workers arrived at this result after studying vibrational and rotational states of hydrogen ($H_2$) molecules in very distant intergalactic clouds. The hydrogen molecule can be likened roughly to a dumbbell, with the two weights at the ends representing the protons and the central bar the molecular link provided by the two shared electrons. The vibrations and rotations of such an object are determined by the ratio of the masses of its constituent protons and electrons.

Certain theories claim that a variation of this parameter could exceed a variation of $\alpha$ by more than 10 times. It should therefore be easier to find, but this has not been the case. Measuring it is in fact more problematical than measuring any variation of $\alpha$, for there are very few clouds that lend themselves to such a measurement! Because the hydrogen molecule is very fragile, any ambient ultraviolet radiation will easily dissociate it. Also, it usually forms only in the presence of dust, inside very dense clouds. Now, if these dense clouds are also very small, they are unlikely to lie across the line of sight of a quasar. A French group led by Cédric Ledoux, based at the European Southern Observatory in Chile, has carried out the only systematic observational studies of molecular hydrogen in the spectra of quasars. They have found six clouds containing these molecules, but only two of these were chosen for measurement, as not all such clouds are suitable for this work. Firstly, the absorption lines must be neither too pronounced nor too weak, but just right for measurement; and, secondly, the structure of the cloud must be simple enough to allow modelling of the lines using the fewest parameters. Finally, the quasar must be bright enough to provide data of acceptable quality. Further observations of these two exceptional clouds will lead to even better measurements.

The first measurements made by the Franco-Indian group seem to show something interesting. They indicate that the ratio $\mu$ has varied by a few hundred-thousandths ($10^{-5}$) over several billion years. Again, we must proceed with caution, and await confirmation of this measurement by other means. There is at present a laboratory investigation of the vibrational and rotational spectra of these molecules, one of the limiting factors in the research into $\mu$. Ultraviolet light is strongly absorbed in air, and there are few UV spectrographs that can obtain the measurements. One of these is at the Meudon Observatory, where most of the measurements have been carried out. In spite of all the efforts of the experimental scientists, the accuracy of the laboratory measurements is less than that obtained from observations of the far reaches of the universe. Surprising as this may seem, it is a fact and it is possible that systematic errors are creeping into the measurements.

Whatever the truth of this may be, there is no denying both the sheer doggedness of the physicists in their quest to check on dimensionless parameters, and their desire to advance our knowledge of physics. Will some of them realise their dream of detecting a variation? Our commission of inquiry also asks this question. What theoretical arguments motivate the researchers? What kind of revolution do they expect? What theoretical paths will they follow? What is at stake if the revolution comes? The time has come to explore some higher spheres of theoretical physics.

# 7

# What if they *did* vary? Gravity waits in the wings

At this stage in our inquiry, no variation in any fundamental parameter has been confirmed, so modern physics has no need to feel threatened. On the contrary, its bases and principles seem to be reinforced by what we have seen in the course of our inquiry. Yet the doubts still linger, and there is a vague sense of disappointment in the air. It seems that, in the final analysis, we almost want there to be such a variation. Physicists, and especially the theoretical kind, seek out the slightest drift of a constant, not in dread but in hope. What are they hoping for? Why do modern theoretical physicists feel unfulfilled?

## FORCES TO BE UNIFIED

The first reason for a sense of dissatisfaction is that scientists still cannot unify gravity and the three other fundamental interactions. From 1915 onwards, having outlined general relativity, Einstein carried out research into the unification of gravity and the only other interaction then known, the electromagnetic force.

What differentiates these two forces? As Dirac pointed out, with his large numbers hypothesis, the strengths of the two forces are not commensurate: when he examined the ratio of the gravitational and electromagnetic forces between the proton and the electron, he came up, as we have already seen, with the enormous number $10^{39}$. At first sight, they have similar influences, since the strength of both diminishes as the inverse square of the distance between the two bodies involved. However, this is modified by one essential characteristic: there are two sorts of electric charge, negative and positive, while gravity has only mass, which is strictly a positive charge. The electromagnetic force is either attractive or repulsive, depending on whether the charges are of the same or opposite signs. Consequently, the influence of a charge may be compensated for by a screening effect due to opposite charges being present in the vicinity. On a large enough scale, matter always has no net charge, meaning that the electromagnetic force effectively no longer acts. On the other hand, the influence of gravity is always there. Also, although its influence is much weaker, it still controls the motions of bodies at very large distances, and leads to enormous accumulations of matter such as stars and galaxies.

There is another difference between the two forces: while all bodies fall in the same way in an external gravitational field, the motion of a charged particle in an electrical field depends on its charge to mass ratio. For example, a singly ionised atom and a doubly ionised atom have almost the same mass, but will not move in the same manner through an electric field. Einstein made use of the universality of free fall in his description of the geometry of gravitation, transforming it into a property of curved space-time. Since all bodies are affected in the same way in a gravitational field, irrespective of their mass or chemical composition, this gravitational field is equivalent locally to acceleration, which can be cancelled out by substituting a reference frame in free fall. Since he could not ascribe this property to the electromagnetic force, the father of general relativity sought another way to transform it, like gravity, into a property of space-time. He pursued this idea for 30 years, without success.

Meanwhile, most physicists were exploring a different path towards the unification of forces, a path that in the end proved more profitable than Einstein's. During the 1940s, the electromagnetic field was 'quantised' within the framework of quantum field theory, the most accomplished and best verified in modern physics. Simultaneously embracing quantum effects (finite Planck constant $h$) and relativistic effects (finite inverse of the speed of light $1/c$), it is found at corner 6 of the cube of physical theories (see Chapter 3). While two extra fundamental interactions – the strong and weak nuclear forces – had been found, the great merit of the quantum field theory was the unification of the electromagnetic and nuclear interactions in the 'electroweak' interaction. In everyday conditions, the two interactions appear different, but at high energies, for example in particle accelerators, they are merely two facets of the same interaction. It was shown that the strength of an interaction depends on the energy involved: forces of different strengths in low-energy conditions may reveal themselves to be of the same strength above a certain energy level. This offers the possibility of describing both forces in a unified manner. To justify the changes in symmetry of a theory when it involves lower energies, theoreticians introduced a mechanism known as 'symmetry breaking', similar to what occurs in a phase transition, for example the transition from a liquid to a crystalline state. With this mechanism, the theoreticians associated a new particle, the 'Higgs boson' named after one of them, Peter Higgs. This particle is at present being sought with the aid of various particle accelerators.

The quest for unification received a boost in the 1960s, when researchers realised that it was possible to unify the strong nuclear interaction and the electroweak interaction. However, gravity could not be accommodated in this scheme, for two reasons: a theory of 'quantum gravitation' (which would occupy corner 7 of the cube of theories) remained elusive; as did another familiar obstacle, the vast difference between the strength of gravity and the three other fundamental interactions. Dirac was right to stress that this disparity would prove to be a hindrance to a unified picture of the interactions, and is known today as the hierarchy problem.

## ENLARGING THE THEORETICAL FRAMEWORK

The quantisation of gravity may have come up against a brick wall, but what of the geometrisation of the other forces? Let us return to the early twentieth century, when a German mathematician of Czech origin, Theodor Kaluza, attempted the geometrisation of the electromagnetic force in 1919, by proposing a generalisation of general relativity. Remember that general relativity links the distribution of matter and energy in the universe to the geometry of space-time, which is identified with the gravitational field. This space-time geometry is described by a tensor, i.e. a set of components arranged in a rectangular table, symmetrically positioned around its diagonal (Figure 7.1). Common sense tells us that we live in a three-dimensional space, and in one-dimensional time. The tensor representing this space-time in four dimensions has four lines and four columns, and has therefore 10 independent components. Kaluza had the original idea that space might not have only three dimensions, but four, which implied a five-dimensional space-time. This extension led to the addition of a fifth column and a fifth line to the space-time tensor, giving it 15 independent components. Kaluza judiciously interpreted the new components of the tensor as belonging to an electromagnetic field and a new scalar field (scalar since it was represented by only one component), not corresponding to any known interaction.

To make sense of such a construct, we need to explain why we are not conscious of this fourth dimension of space. A decade after Kaluza's idea, Swedish physicist Oskar Klein suggested that this dimension curls in upon itself,

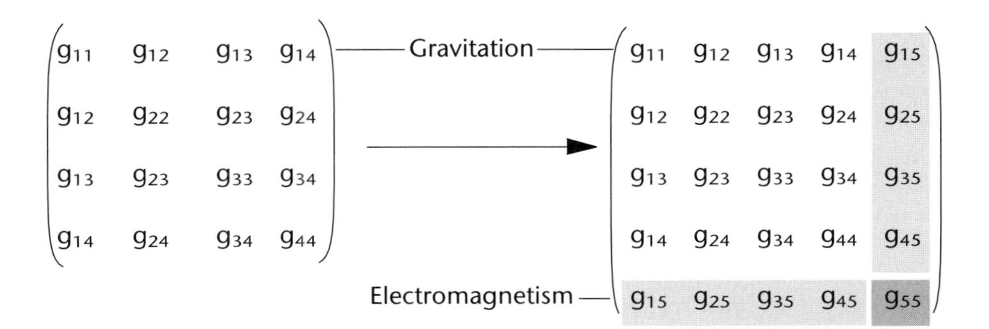


Figure 7.1 Metric tensor relating the curvature of the universe to its matter and energy content, in the case of general relativity (left), and in Theodor Kaluza's theory (right). Kaluza added a dimension to space, effectively adding an extra line and column to the tensor. Their components correspond to the electromagnetic field (light grey) and a scalar field (dark grey).

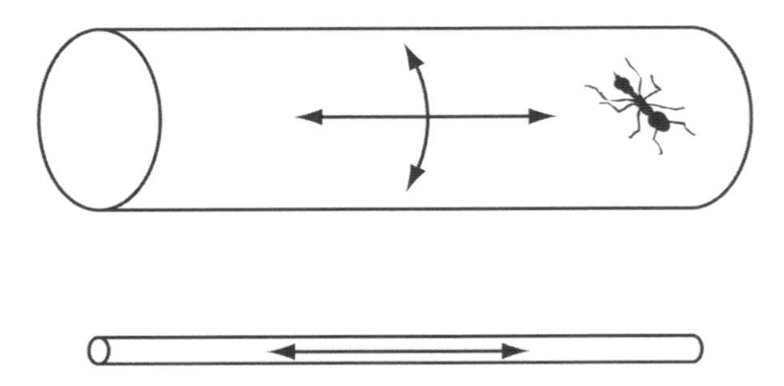

Figure 7.2 Let us reduce the world to a cylinder. If we could, like an ant on a cable, 'measure' our position using more than just the circumference of the cylinder, we would perceive that we are in a two-dimensional space (we need two coordinates to work out our position). If not, we have the impression that we are in a one-dimensional space, the second dimension having been 'compactified'.

with such a small radius that it is not amenable to measurement with our instruments. Imagine that you are a tightrope walker on a wire: your foot is too wide to feel the width of the wire, giving you the sensation that it is one-dimensional. On the other hand, this wire would seem like a cylinder to an ant that is moving across the wire's surface to avoid being squashed by your foot! In a way, the ant can be seen as a measuring instrument capable of higher resolution than your foot, since the whole structure of the wire is accessible to its scrutiny (Figure 7.2). Similarly, we can conceive the physical existence of an extra dimension, curled into a cylinder, with a diameter below the resolution of our measuring instruments. The diameter of this cylinder, also known as the scale of *compactification*, is so microscopic as to be inaccessible to us, and we therefore do not experience this particular dimension.

This 'Kaluza–Klein' theory met with little acceptance, and Einstein, to whom Kaluza sent his article, found it interesting, but concluded that it was ultimately unnecessary. However, it inspired many attempts at the unification of fundamental forces, and the idea of compact extra dimensions has had a long run!

Dirac's large numbers hypothesis opened up another avenue of approach. We recall his argument that the ratio of the electromagnetic and gravitational forces between the electron and the proton increases in proportion to the age of the universe. If so, the strengths of the two forces would have been comparable at the beginning of history and the hierarchy problem would simply be a consequence of the great age of our universe. Dirac's hypothesis implies that at least one of the characteristic constants of these forces must vary with time. The offender, he said, was the gravitational constant, diminishing in inverse proportion to the age of the universe. In itself, this argument is not a physical theory, but it does make use of a numerical coincidence to point the way towards the formulation of such a theory. Note also that the gravitational constant $G$ is dimensional, while it is dimensionless parameters that retain a physical sense. In reality, Dirac considered the dimensionless ratio $Gm_p^2/hc$ and declared that the mass of particles, Planck's constant $h$ and the speed of light $c$ remain constant.

## DIRAC'S OFFSPRING

Simple astrophysical considerations soon put paid to the hypothesis of a variation of the gravitational constant. In 1948, Edward Teller, a Hungarian-born American physicist, stated that, if the gravitational constant had varied with time, then the Sun would have been more luminous in the past and the surface of the Earth would have been hotter. He calculated that the oceans would have reached boiling point in the Cambrian period, about 500 million years ago, and life would never have developed there, contrary to the fossil evidence of organisms such as algae and trilobites. (In reality, Teller estimated that the boiling of the oceans would have occurred some 200–300 million years ago, the age of the universe at the time having been underestimated. Then, life was more complex, and the first tetrapods were emerging from the water.) Other physicists, such as Gamow, argued that, if the Sun had been more luminous in the past, it would have consumed its reserves of hydrogen more quickly, and would by now have exhausted its hydrogen and become a red giant, contrary to all observational evidence! More recent measurements have shown that the gravitational constant $G$ cannot have varied by more than 1 part in 10 billion! Such a minuscule variation could have made itself apparent only over periods of the order of the age of the universe.

Even in the article in which he criticised it, Gamow wrote that it "would be a pity to abandon to abandon an idea so attractive and elegant as Dirac's proposal" – especially since the analyses set out to refute it lacked theoretical consistency, as German physicist Pascual Jordan showed in 1949. Jordan saw as their main fault their reliance upon physical laws that were derived from the assumption that $G$ was a constant while later making it vary in proportion to time. Dirac foresaw that this may well be a valid approximation if the variation of $G$ is slow. Strictly speaking, however, we cannot be content with approximation: if we assume that the gravitational constant varies with time, we must profoundly revise the physical laws that accept it. We must construct a new theory of gravity: an extension of general relativity which encompasses it just as general relativity encompasses the Newtonian theory of gravity. First and foremost, this theory would have to succeed where the theory of general relativity succeeds. It would also have to provide a crucial element: the equation of the evolution of $G$. What Dirac postulated was an inverse law of the age of the universe, but there is nothing to prove that this choice is compatible with the required theory.

In his quest for this new theory, Jordan tried to promote the gravitational constant to the rank of a dynamical variable. He replaced it with a scalar field, characterised by a specific value at each point of space at each moment. So, with its value being localised and instantaneous, $G$ lost its status as a universal constant. As for the new scalar field, it was deemed to be responsible for a new, uncompensated and long-range interaction, which made it very similar to gravitation.

Swiss physicist Markus Fierz took Jordan's work further, and showed especially that other constants – for example, the fine-structure constant – might vary under certain conditions concerning the interaction of the new scalar field with ordinary matter. If this were true, he said, atomic spectra would differ from one point of space-time to another.

In 1961, Robert Dicke and his student Carl Brans worked these ideas into a theory of gravity within which $G$ can vary: a theory encompassing general relativity. Several similar theories that have been developed are known collectively as 'tensor–scalar' theories of gravity. Some of these liken the additional scalar field to that which appears in the five-dimensional tensor of the Kaluza–Klein theory, which is one way to place the pieces of the jigsaw. For all these nascent theories, any variation of the gravitational constant $G$ would be an important encouragement. The variation of another fundamental constant would also have important consequences...

## GENERAL RELATIVITY: TRIAL BY CONSTANTS

In 1967, Gamow set out to test Dirac's "attractive and elegant idea" by turning his focus on the fine-structure constant. It is true that, within the framework of the large numbers hypothesis, two constants may be assumed to be variable: either $G$ decreases – as Dirac suggested – or $e^2$ increases in proportion to the age of the universe. This second solution has a definite advantage, since the fine-structure constant is proportional to $e^2$. These two possibilities lead on to distinct theories, with different observable predictions. In fact, a variation of $G$ alone would not affect atomic spectra and would involve only the gravitational sector. On the other hand, a variation of the fine-structure constant would involve a modification of the spectra. Consequently, we could distinguish between the two hypotheses through observation.

As we have seen, a theory within which $G$ varied could encompass general relativity. Conversely, a definite variation of a fundamental constant, be it $G$ or $\alpha$ or $\mu$ (the ratio of the masses of the proton and electron), would call into question general relativity itself, by directly undermining its central pillar, Einstein's equivalence principle. This principle incorporates three postulates. The first states that the outcome of any local non-gravitational experiment is independent of where and when in the universe it is performed. This is known as local position invariance. The second (local Lorentz invariance) states that the outcome of any local non-gravitational experiment is independent of the velocity of the freely-falling reference frame in which it is performed. The third deals with the universality of free fall: two different test bodies in an external gravitational field fall with the same acceleration, independent of their masses or their chemical composition. General relativity, and indeed tensor–scalar theories, claim that if these three postulates are respected, gravity becomes a simple consequence of the geometry of space-time. It should be remembered that Einstein's equivalence principle includes only non-gravitational experiments, i.e. those in which gravitational binding energy is negligible, as in spectroscopic experiments. If we include gravitational experiments, we encounter the strong equivalence principle, and the theory of general relativity is thought to be the only theory of gravity that satisfies this principle.

Let us now see how the variation of a fundamental constant other than $G$ would go against the principle of equivalence. We can say immediately that if such a constant varied in time or space: certain experiments would not be reproducible; the hypothesis of global position invariance would be violated; Einstein's equivalence principle would fall, with negative implications for general relativity; and gravity could no longer be seen as a geometrical phenomenon. Therefore, testing the stability of constants amounts to testing general relativity itself: it is a test of fundamental physics.

Through slightly more elaborate reasoning, we shall show that the variation of a fundamental constant would imply a violation of the universality of free fall. It is known that the mass of any body is given by the sum of the masses of its constituents, to which must be added the binding energy ensuring its cohesion, which is obtained from the equation $E = mc^2$. Let us assume that $\alpha$ varies in space. This means that the electromagnetic energy responsible for cohesion in bodies will vary from one place to another. It would then follow that the mass of an object would depend on where it was located. According to the principle of the conservation of energy, a body whose mass varies from one point of its trajectory to another would accelerate when moving towards a region where its mass is lower, and decelerate when moving towards a region where its mass is higher. Moreover, since the electromagnetic energy of the proton differs from that of the neutron, this variation in mass would depend on the relative numbers of protons and neutrons in an object, i.e. on its chemical composition. In other words, bodies in free fall in space would experience a new force, dependent upon their composition. This result would contradict the idea of the universality of free fall. The same kind of reasoning can be used in the case of $\mu$. So the variation of one of these constants would signal a crack in Einstein's principle of equivalence, and underline the need for a theory of gravity that would extend the framework of general relativity. This theory would have to introduce a new force, working at long range and dependent on an object's chemical composition. This force would be associated with a new field, also responsible for the variation in the 'offending' constant, as already shown in the work of Jordan and Fierz.

Conversely, studies on the universality of free fall give us information about the possible variation of constants. In the laboratory, we use clocks and torsion balances to achieve very accurate results: the universality of free fall has been verified to an accuracy of one part in a trillion ($10^{-12}$) between masses of copper and beryllium, and 1 part in 10 trillion ($10^{-13}$) between rocks from the lunar mantle and samples from the Earth's core.

In these two experiments, binding gravitational energy is negligible. Other experiments have involved the investigation of the relative 'falls' of the Earth and the Moon in the gravitational field of the Sun. If these two bodies were not subject to the same acceleration, their trajectories would differ from those calculated. Here, since the gravitational energy between the planet and its satellite is not negligible, we are testing the strong equivalence principle. For three decades now, astronomers have been measuring the distance from the Earth to the Moon by means of 'laser ranging'. Laser pulses are emitted from four Earth stations (Haleakala in Hawaii, Wettzel in Austria, Grasse (near Nice in France), and McDonald, Texas, USA – see Figure 7.3). The light is reflected back to

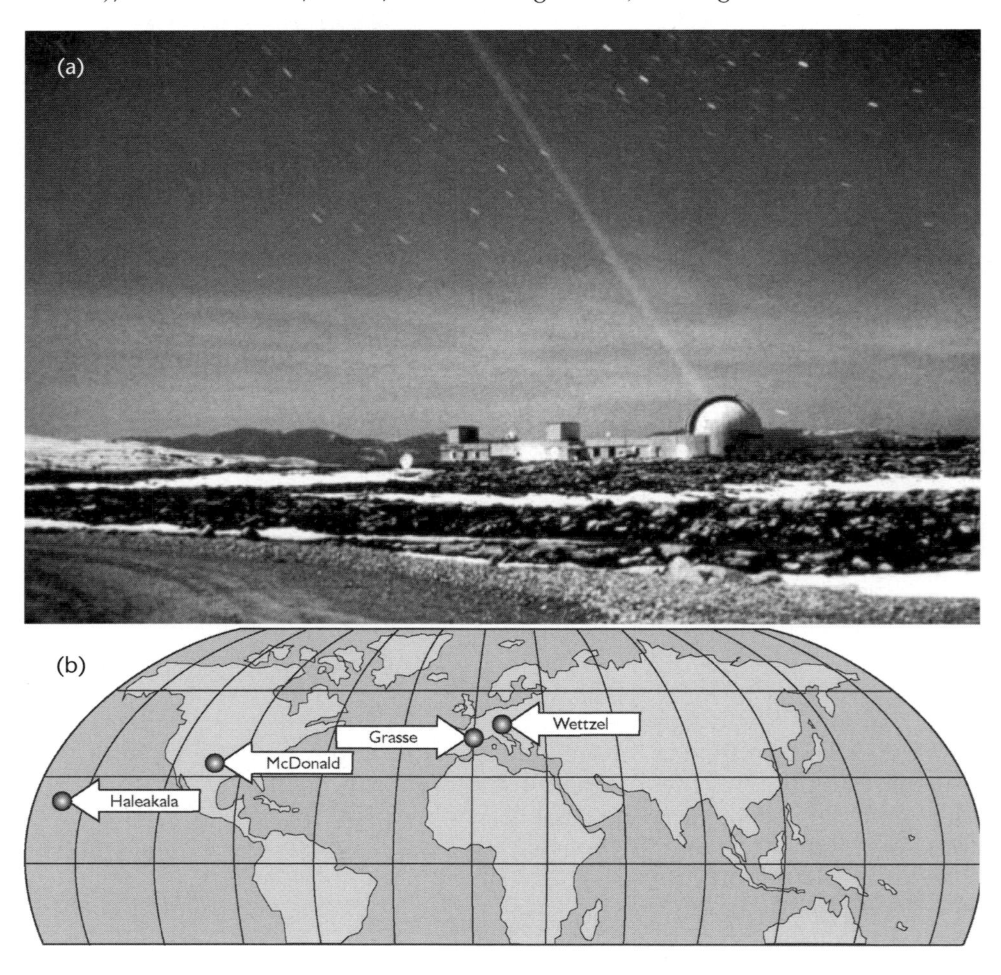


Figure 7.3 Lunar laser ranging (I). (a) A laser pulse is emitted from the Grasse Observatory (© Côte d'Azur Observatory). (b) The four laser ranging stations measuring the Earth–Moon distance.

(a)

(b)

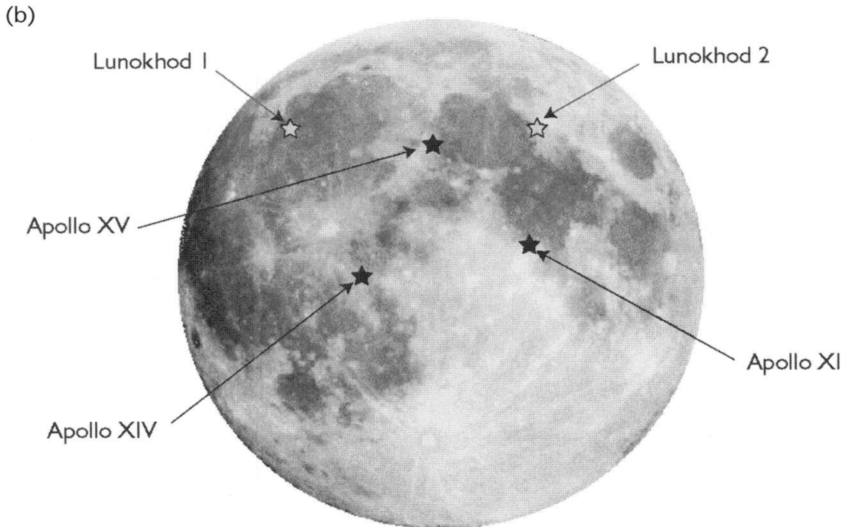


Figure 7.4 Lunar laser ranging (II). (a) Reflector left on the Moon by the Apollo 11 mission (© NASA). (b) Five reflector sites on the Moon (© NASA).

Earth by mirrors taken to the Moon by the American Apollo astronauts and the Franco-Russian Lunokhod mission (Figure 7.4). The time taken by the light to travel to the Moon and return to Earth gives a measurement of the distance of the Moon to within 1 centimetre, and these measurements agree completely with the predictions of general relativity. To date, the equivalence principle has been verified with great accuracy, to just a few parts in 10 trillion ($10^{-13}$).

From these experiments we can deduce a limit for the spatial variation of the fine-structure constant: over a distance of the order of the radius of the Earth's orbit around the Sun (150 million kilometres), the spatial variation of $\alpha$ remains less than 1 part in ($10^{-32}$) per centimetre. Roughly extrapolated to the size of the observable universe (15 billion light-years), assuming it to be linear, the variation of $\alpha$ can be no more than 1 part in 10 thousand ($10^{-4}$). Incidentally, the variation announced by Webb and his colleagues would be 10 times smaller.

Even more accurate experiments are being planned. Two space experiments, MICROSCOPE (Figure 7.5) and NASA's STEP, should increase the accuracy of the estimation of $\alpha$ by between 100 and 100 000 times. The European Space Agency hopes to launch MICROSCOPE in 2008. If some deviation from the universality of free fall, however minuscule, were found, or any variation in the fine-structure constant, we would then have irrefutable proof of the need for new standards in physics. It is a special attribute of physicists that they are not easily discouraged. If, without further inquiry, they had taken on board Galileo's and Descartes' findings about the propagation of light, they would never have measured its speed!

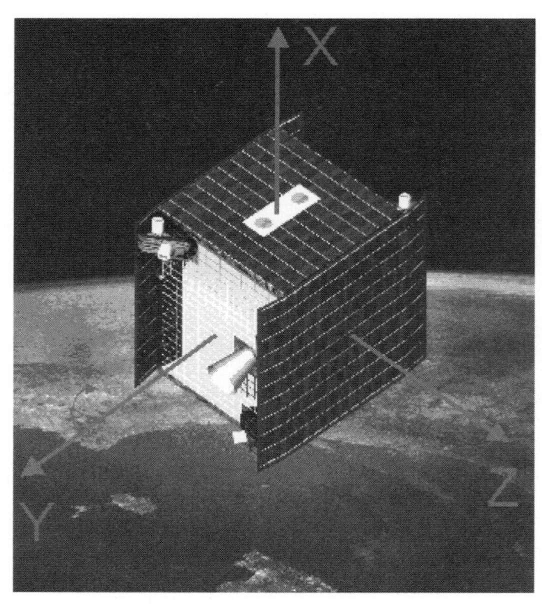


Figure 7.5 The CNES satellite that will carry the MICROSCOPE experiment, comparing the free fall of two bodies of different chemical compositions. (© CNES.)

## DARK IDEAS FOR COSMOLOGISTS

As we have already mentioned, the unification of theories describing fundamental interactions is the deep- seated motive that drives theoreticians to pursue a more global theory than general relativity to describe gravity. One question haunts physicists and touches on the universality of the existing theories: is it legitimate to apply physical theories deduced from present-day, local observations to the distant past, and to places far removed? General relativity has been tested out within the solar system and in the rest of the Milky Way, but are we, like the greenfly on the rose bush on a summer's day, imagining it to be the whole world, while certain concepts - the changing seasons and climate, areas where no roses grow - are hidden from us?

Is gravity properly described by general relativity throughout the universe? Cosmological observations justify asking such a question. The object of cosmology is to study and map the universe as a whole. Gravity rules the cosmos: with a long range that can never be cancelled out by any screening effect; it is the force that structures the large-scale universe. But does it do this entirely according to the predictions of general relativity?

Galaxies, the fundamental entities of the cosmos, have been the subject of a host of studies: their shapes, composition, ages, dynamics and other aspects are all investigated by detection of the electromagnetic radiations they emit. Visible light, radio waves, and X-rays all give an account of what is known as *visible matter*, but astronomers cannot fully explain the internal dynamics of galaxies through observations of this matter alone. Recalling that a deviation of Uranus from the orbit predicted by Newton's theory led to the discovery of Neptune by Le Verrier and other astronomers, we see a similar story with the dynamics of galaxies: they do not comply with the predictions of general relativity. In particular, stars in the outer regions of spiral galaxies orbit more quickly around the centre of these galaxies than the theory predicts, assuming of course that these galaxies are composed only of visible matter. These observations lead astrophysicists to the conclusion that the galaxies contain more matter than is apparent. What sets this apart from the discovery of Neptune is that this 'dark' or 'hidden' matter, invisible to our instruments, would have to be ten times more abundant than visible matter to cause such effects. Therefore, we are not just dealing with a simple correction!

What might the nature of this dark matter be? To date, this question remains unanswered. Is it 'ordinary' baryonic matter, like the matter that surrounds us? Is it lurking unseen in vast clouds or haloes, too cold and tenuous to emit radiation? Or is it in the form of 'brown dwarfs', objects halfway between planets and stars and therefore not easily observed? Neither absorption spectra of cool clouds, nor recent observing programmes targeting brown dwarfs, have come up with the required populations. Furthermore, predictions based on primordial nucleosynthesis have established that baryonic matter, composed essentially of protons and neutrons, represents only 4 per cent of the total matter of the universe. It therefore cannot explain the nature of all dark matter.

Dark matter must be different in its nature, and many candidate constituents have been put forward from among the elementary particles: for example the neutrino, which displays almost no interaction with matter or light, and is therefore difficult to detect. The mass of neutrinos was for a long time considered to be zero, eliminating them from the list of candidates. Then, in the early 2000s, certain observations of solar neutrinos, together with other experiments, indicated that the mass of at least one of the three types of neutrino might be non-zero. However, the value is still too small to account for all the dark matter, so perhaps it comprises other massive particles as yet unknown, and just as difficult to capture as neutrinos.

Or perhaps... perhaps the world of the greenfly isn't the same outside their garden. The conclusion on the existence of dark matter depends on one crucial hypothesis: the validity of general relativity to describe gravity, since dark matter is revealed though gravitational effects. The alternative is to question general relativity.

## THE COSMOLOGICAL CONSTANT & CO.

Dark matter is not the only 'anomaly' that arises out of our observations of the large-scale universe. In the years following 2000, cosmologists realised that matter, visible or dark, was not able to explain the dynamic of the whole universe! The universe is expanding, as Hubble's fleeing galaxies indicate. Because of the long-range influence of gravity, this expansion should slow down as the universe grows larger. To be more precise, there is a theorem that states that, if general relativity correctly describes gravity and if the universe is composed only of ordinary matter, the universal expansion must slow down. But this is not the case: on the contrary, it seems to be accelerating. Among many observations, those made since 1998 of distant type-Ia supernovae seem to be the most convincing. These supernovae are white dwarf stars at the end of their lives, exploding after having accreted too much matter from their environment. Astrophysicists refer to them as 'standard candles', since they understand the characteristics of their explosions well enough, they believe, to be able to determine their distances from their luminosities. Now the distances of these remote type-Ia supernovae, as calculated from their luminosities, are not in agreement with the distances suggested by their spectral signatures (assuming a decelerating cosmic expansion). They all seem systematically further away than they should be. There is only one possible explanation: at some time in the last few billion years, the expansion of the universe began to accelerate. But how can this acceleration be accounted for?

Manifestly, we must call into question one of the two hypotheses of the theorem that predicts a deceleration of the expansion. Either

- general relativity is correct, but the universe is bathed in some exotic matter (over and above dark matter!) that is gravitationally repulsive; or

- general relativity does not correctly describe gravity on the large cosmological scales, in which case it must be replaced by a wider theory that will still be compatible with the predictions of general relativity within the solar system.

Whatever the case, a new kind of physics looks likely for the new century: just tinkering with the old theories will not suffice. The theoretical reflections of Dirac and Jordan still echo today.

A provisional solution to the problem of the acceleration of the universe would be to reintegrate the *cosmological constant* into the equation of general relativity. Einstein introduced this constant in the first cosmological model deduced from general relativity in 1917. With this model, he aimed to establish a static, closed universe: the cosmological constant would balance out the effects of gravity and prevent the static universe from collapsing in upon itself. Ever since, this constant has been going in and out of fashion among cosmologists as new observations are made. It still remains the ideal candidate for explaining the acceleration of the universe, but its theoretical significance is far from being understood.

Theoretical physicists have identified this cosmological constant with so-called 'vacuum energy'. According to quantum field theory, a vacuum (empty space) possesses energy. Now general relativity tells us that all forms of energy create gravitational fields, therefore the energy of the vacuum can contribute to the expansion of the universe. Unfortunately, a problem of scale has worried the researchers: the value of this kind of energy is immensely greater – in fact, $10^{120}$ times greater (!) – than the value deduced from cosmological observations. The gravitational effect of this vacuum energy, deduced from quantum field theory and particle physics, would radically change the look of our universe: for example, galaxies could never have formed! Yet another problem to add to our gravitational checklist, and one which highlights one of the limitations of general relativity: this theory links the matter present in the universe (in quantum terms) with the properties of space-time (dealt with in classical or non-quantum terms). The cosmological constant problem is perhaps another thread leading us through the labyrinth towards a quantum theory of gravity. Right now, it serves only to underline the great need for such a theory.

As well as the cosmological constant, many other candidates for exotic matter have been proposed, but none has yet been detected in either laboratories or accelerators. These possibilities are collectively and simply known as 'dark energy'.

Whatever the answer is to the enigma of the accelerated expansion of the universe (cosmological constant? vacuum energy? dark energy? a new theory of gravity?) it could bring with it consequences greater than those associated with the enigma of the black body in the late nineteenth century. Finding that answer will either reveal the existence of an as yet unknown type of matter, or point the way to a better understanding of gravity. Physicists are enthusiastically exploring this *terra incognita*, bringing to bear all the means at their disposal. They carry out an ever-increasing tally of observations, both of the cosmos and in particle accelerators. They test gravity within the solar system, and work with those unavoidable fundamental constants.

Our inquiry has now revealed some of the difficulties that modern physics has encountered. Firstly, not all interactions are described in comparable frameworks, which suggests that those frameworks are too narrow. Secondly, the laws of gravity do not adequately describe the dynamics of galaxies, unless there is some new, dark matter out there that is invisible to our instruments. Lastly, the recent dynamics of the universe as a whole is not yet understood.

Our commission of inquiry agrees with the physicists: we must uncover the flaw in today's physics, in order to pursue a new and more general theory that will overcome these difficulties. Now, the study of fundamental constants offers a clue to the nature of this flaw. A variation in certain constants would indicate that we must extend general relativity and construct a new theory of gravity. If such a variation were proved, to which theory would we turn? How would our world view be affected? These questions usher in the last stage of our inquiry.

# 8

# Strings attached: towards multiple universes

Our commission well understands the fascination that the study of constants holds for many physicists. At stake in this research is a theory that unifies interactions and is applicable to all scales of our universe. By modifying our knowledge of the world, this new theory would probably alter our list of fundamental constants. Can we imagine how? This time, the inquiry is more forward-looking and speculative...

## THE HARMONY OF THE WORLD

Among the theories put forward to try to resolve the various problems mentioned in the previous chapter, there is one very serious candidate: string (or superstring) theory. String theory replaces the concept of fields, which generate fundamental particles within the quantum field theory, with the addition of objects known as 'strings', by which one identifies the vibratory modes of the fundamental particles.

The seed of this idea was planted in the 1960s. Gabriele Veneziano, a talented Italian physicist, was investigating the characteristics of the strong nuclear interaction. Unlike the electrical field, which extends throughout space, the strong-interaction field between two quarks is concentrated into a narrow tube between them. The image of this tube gave Veneziano the idea of a linear object along which energy spreads: a particle was no longer described by a point, but a line. The act of replacing a single point with an infinity of points on a line gives access to an infinity of states, within which can be described an infinity of particles. Piano and violin strings emit a variety of different notes, according to their modes of vibration. Similarly, the vibrational states of a fundamental *string*, characterised by its tension, determine the masses and spins of a large number of particles. It is therefore possible to reduce many parameters to only a few particular aspects of a single theoretical entity: the string.

In 1974, theoreticians discovered that this economical image of a fundamental vibrating string might also describe gravity. The reason is that, among the states of excitation of the string, there is a particle whose characteristics – for example, mass and spin – make it a natural mediator of the force of gravity. This particle is known as the graviton. String theory therefore included Einstein's general relativity in a coherent quantum framework. It was also capable of unifying all the fundamental interactions, including the quantum field theory and general relativity. Could string theory occupy corner 8 in the cube of theories? It then looked simple on paper, but string theory has now far too many loose ends. However, for the purposes of our inquiry into the edifice of physics, we can retain two crucial aspects of it.

In the first place, string theory offers a new perspective on the relationship between space-time, gravity and matter. In the framework of general relativity, space-time contains matter, which acts upon its geometry. In the framework of string theory, the structure of space-time flows from one of the characteristics of matter itself, since the fundamental string naturally contains the graviton. Hence, this new theory upsets the relationship between the container (space-time) and the contents (matter).

The second contribution of string theory touches more directly upon the subject of this book: the role of physical constants.

## HOW MANY PARAMETERS ARE UNDETERMINED?

Let us spend some time recalling Dirac's initial questions about the parameters of physics. Why is the value of the fine-structure constant 1/137, and not 1/256 or some other number? Why is the ratio of the masses of the proton and the electron 1,836, and not 10 times less, or 10 times greater? Although physics is assumed to explain the world, it still could not justify these values – a fact that caused Dirac a great deal of dissatisfaction!

Today, the situation looks no more promising than in Dirac's day. The standard model of particle physics admits 18 arbitrary parameters, including the masses of fundamental particles and the characteristics of the four interactions. Add three (dimensional) constants chosen as fundamental units, and we have 21 undetermined constants. The number rises to 25 if the three types of neutrino that exist in nature have mass. It could even reach almost 100 in certain extensions of the standard model of particle physics. These constants, however, are considered to be arbitrary, as they could have had different values without threatening our theoretical edifice. The world would be different, but not the physical theories. Thus, if the fine-structure constant had a value that was twice as large, the edifice of particle physics would not be in danger, even though chemistry and biology would probably be disrupted.

*A priori*, if string theory turns out to be relevant, it will shorten the list of fundamental constants. The only indispensable ones would be the tension $T$ of the fundamental string, which determines its fundamental modes, the string coupling constant $g_s$, and the speed of light $c$. And some fundamental parameters could certainly disappear from the list of constants. In 1987, American mathematician and physicist Edward Witten remarked that all dimensionless constants, or fundamental parameters, would become dynamical quantities within the framework of string theory. This property offers the hope of developing a physics of constants (or 'constantology' – a term we coined earlier in this book), and it could lead to a better understanding of why constants have their observed values. Alternatively, if these parameters have been able to vary during the history of the universe, investigations into their variation could offer the first observational clue to support string theory.

Will we be able, in the course of time, to deduce from string theory all the fundamental parameters of the standard model of particle physics, by calculating them as a function of the coupling constant $g_s$? The answer is far from clear. In fact, even if the equations depend on only one parameter, their solutions may depend on several others. To take an analogy: this is already the case in Newton's theory, which predicts the elliptical shape of planetary orbits as Kepler described them, but does not predict the parameters of these orbits, such as radius, ellipticity, etc. These parameters depend on the initial conditions in the solar nebula from which the solar system formed. So, no sensible physicist can nowadays play with the idea of calculating the characteristics of these orbits *ab initio*. The best one could do is to calculate the statistical distribution of these characteristics. Similarly, string theory leaves some parameters undetermined. But which ones?

For mathematical reasons, string theory considers space to have nine dimensions, i.e. six more than our usual three. Like the fifth dimension of the Kaluza–Klein models, these extra dimensions of space are 'compactified' within distances smaller than the most powerful particle accelerators can survey. If space possessed only one extra dimension, its radius of compactification could be seen as an extra parameter that would determine the solutions to the theory. The situation is more complicated if we have to describe a six-dimensional space, for then we would have hundreds of parameters to choose from. Each set of parameters involves a solution to string theory, and a universe with its own particular physical characteristics. In summary, the parameters of the standard model of particle physics depend (certainly) upon the string coupling constant, and also upon numerous geometrical parameters.

Since the theory does not predict their values, can we hope to determine them by experiment? This seems difficult. Let us suppose that the radii of compactification of the extra dimensions are of the order of a micrometre (one thousandth of a millimetre), itself very large compared with the scales encountered in particle physics. This means that, over distances of less than a micrometre, gravity could behave differently from the predictions of general relativity, or rather Newton's theory, which are quite correct at these scales. Nowadays, Newton's law is tested down to distances of 300 micrometres, but, below this, no investigations have yet succeeded, since many other forces then overcome the gravitational force: for example, the Van der Waals forces, of electromagnetic origin, or the (quantum) Casimir force. Experimental physicists are trying to push down the limits as far as possible in the hope of revealing the existence of these extra dimensions.

## THE OBSERVER: A NECESSARY CONDITION

String theory can take us into universes with very different physical properties, but how do we explain the fact that our universe is the way it is? Why do constants have the particular values ascribed to them? These values seem to be very precisely laid down. Is this an inevitable necessity; a finality? Or is there some mechanism capable of setting these values?

A 'finalistic' approach is one that supposes that the appearance of life, or of humankind, in the universe must necessarily be a product of some design. As such, it is outside the framework of science, whose methodology rejects the idea of a final cause; and it is also, for the same reason, outside the remit of our inquiry.

A second approach entails one of the considerations originally developed by Robert Dicke in 1961. Commenting on Dirac's remarks on the coincidence of large numbers, Dicke sought to justify the current value of the age of the universe. He noted that it must be old enough for first-generation stars to have formed and to have produced the chemical elements, like carbon, of which living things are made. On the other hand, if the universe were considerably older, stars would by now have collapsed into white dwarfs, neutron stars or black holes, and life as we know it could not exist. Consequently, observers able to examine Dirac's coincidence can exist only if the age of the universe is very close to the observed value.

This observation opened the way to further arguments of this type, where the values of certain dimensionless parameters were imagined to be different: what then would be their effects upon certain physical, chemical or biological phenomena? This kind of argument does not supply an explanation for the values of the parameters, but demonstrates that those values constitute conditions *necessary* for the existence of a particular phenomenon: the existence of phenomenon P implies that condition C must be satisfied. It follows that, if C is not satisfied, P cannot occur. However, there is nothing to say that P must necessarily occur. In other words, we show that condition C is necessary, but we do no know if it is 'sufficient'. This argument is of scientific interest, because it defines the conditions under which phenomenon P can occur. In particular, it fixes the value of certain fundamental parameters, when the theory allows their variation.

To help us to understand this approach, let us consider a precise example. Life as we know it appeared through the presence of the element carbon, which acts as a 'skeleton' for long molecules like DNA and proteins. Dicke once humorously remarked that "it is well known that carbon is required to make physicists". Carbon is produced in stars as a result of several nuclear reactions. In the initial stage, two helium nuclei (each composed of two protons and two neutrons) fuse to become a nucleus of beryllium-8. Then this unstable nucleus either decays or fuses with a third helium nucleus to produce a (stable) carbon-12 nucleus. Three conditions are necessary for this process to take place. First, the beryllium must live long enough to interact with another helium nucleus and fuse with it. (Its millionth-of-a-second half-life fulfils this condition.) Next, through a resonance effect, energy conditions must optimise the production of carbon-12 in the star. Using this argument, Fred Hoyle predicted the existence of an excited state of carbon-12, at an energy level equal to the sum of the energies of helium and beryllium, plus an amount of energy provided by the heat of the star. The final necessary condition for the existence of large quantities of carbon is that the carbon-12 nucleus must not suffer the same fate as the beryllium nucleus. If it were to fuse rapidly with another helium nucleus, it would disappear and oxygen would result; but this does not happen because the excited state of oxygen is situated at too low a level to allow a resonance of the type that occurs in the case of carbon. These three conditions exist thanks to a fine equilibrium between the electromagnetic and the strong nuclear forces. In conclusion, the values of the parameters of these interactions constitute a necessary condition for the appearance of life, and hence the existence of humans as observers.

Researching the conditions necessary for the existence of a given phenomenon therefore represents an alternative to the explanation of the values of the parameters. From the beginning of some characteristic phenomenon of our universe, we can define the ranges of the permitted possible values of the parameters of physics. The characteristic phenomenon could be, say, the emergence of human beings (the argument leading to the so-called anthropic principle), or of elementary life-forms (the biotic principle) or of its essential constituents (the carbon principle). It must again be stressed that this reasoning is in no way 'final'; it does not claim, for example, that the universe *must* evolve towards the emergence of humans. Such a proposition would assume a 'strong anthropic principle'. In contrast, the 'weak anthropic principle' considers the emergence of humans as only one among many possible outcomes for a universe. Only those universes in which observers eventually appear can be observed, so that an observer cannot measure *all* possible values of fundamental constants. This approach means that parameters are not explained, but their values are justified *a posteriori*.

## CHOOSING THE RIGHT UNIVERSE

The anthropic principle offers a selectionist vision, similar to that of naturalist Charles Darwin. The notion of mutation driving the appearance of new species would be replaced by the generation of multiple universes, each with its own physical characteristics. The process of selection would be replaced by the possibility of a universe engendering observers capable of observing the values of its parameters. The existence of an observer is therefore a condition that would justify the observed value of certain quantities or the existence of certain numerical coincidences.

To complete this concept, one point remains to be discussed: What becomes of universes conceived through theoretical reasoning with other values for their parameters? If only one universe had been created, how can we understand that it could be realised with values for its fundamental constants leading to the appearance of life? String theory aside, there is a second theory within which we might conceive the existence of universes with different values for their fundamental parameters. This is 'inflation theory', which was first mooted in 1981. Inflation theory supposes that the evolution of the universe has passed through an almost exponential phase of expansion, known as inflation. The predictions of this theory are in agreement with the vast majority of modern cosmological observations. Importantly, it explains the origin of the large-scale structures observed in the universe: galaxies and clusters of galaxies. This theory offers a radically different picture of the universe. Firstly, it is much vaster than just the observable universe. Also, the (eternal) expansion creates regions with their own physical characteristics, as if each had evolved from its own specific Big Bang. Each 'island universe' loses all causal contact with the rest of the universe, of which the observable universe is only an infinitesimal part (Figure 8.1). Thus, all the island universes may acquire their own physical laws, and have different values for fundamental parameters. Our universe represents just one region of one of the innumerable islands created by inflation.

Inflation theory is compatible with string theory: each universe, with its characteristic structure of additional spatial dimensions, can be considered as an island universe. Inflation then becomes a mechanism by which to explore all possible solutions of string theory. It remains to be proved that among all those solutions there is one that corresponds to the physics we observe in our universe. It is such a task that the phrase 'needle in a haystack' comes to mind.

This island universe approach stands distinct from the reductionist quest for an ultimate theory such as Dirac and Eddington envisaged – a theory that could

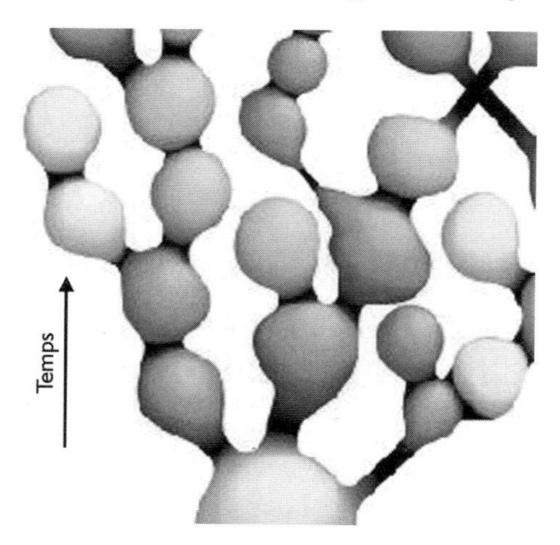


Figure 8.1 In the eternal inflation model, the universe continually produces 'island-universes', and our observable universe would be but an infinitesimal part of one of them. Each island-universe has different values for its fundamental constants. (After Andreï Linde, *Pour la Science*, January 1995.)

explain the values of all the fundamental parameters and make them a quasi-mathematical necessity. What it offers is a theoretical and metaphysical framework that is sufficient to justify our concept that the universe exists as we observe it. If this were not the case, we could not reply to the question: Why is our universe the way it is? On the other hand, we can ask: Does a universe the same as ours exist in the *ensemble* of possible universes within a given theoretical framework, for example string theory? And there is another, more ambitious question: To what extent is our universe generic?

Here, we are at the limits of our investigation into the origin of constants. It must be underlined again that the existence of such a 'multiverse' cannot be tested experimentally!

The notion of multiple universes arises, however, when we take seriously the physical theories we have discovered; this is a notion that is now receiving much attention.

## FINAL REPORT

Our inquiry is nearing its end. The commission of experts has in its files many conclusions on the nature of physical laws and the role of the constants within them. Constants appear unbidden within physical laws, as soon as they are expressed with reference to a coherent system of units. It can be said, however, that they represent the limits of explanation of the theoretical framework in which they appear. Consequently, they represent a real task for those modern physicists trying to explain them or reduce their number.

It is within dimensionless parameters that physical meaning lies. They express the ratios of quantities and are independent of all systems of units. The constancy in time and space of these parameters of Nature flows from the universality of the laws of physics. If observation contradicted this constancy, it would at the same time cast doubt upon the laws of physics. In other words, the quest for a variation in the dimensionless parameters of Nature is no more than a fundamental test of physics itself. In particular, it calls into question general relativity and its geometrical description of gravity. The goal is not, however, to disprove all that has been established, but to move physics forward. Theoretical arguments underline the need for a more fundamental theory of gravity than general relativity offers: arguments such as the search for a unified description of fundamental interactions, and observational arguments about, for example, the dynamics of galaxies or the accelerated expansion of the universe. Any variation of a dimensionless parameter could point physicists along the road to some new theory, and this may enable scientists to put the finishing touches to the structure of modern physics.

## HUNGRY PHYSICS

Our commission has noticed one particular attribute of physics: the theories it develops tend, historically speaking, to absorb external entities in order to resolve certain problems. Let us explain. In Newtonian physics, space, time and constants do not require an explanation: they are immutable entities laid down by Nature. They define the 'stage' upon which physics unfolds. In this sense, they are outside the realm of an explanation for the theory. It was Einstein who overthrew this idea when he incorporated space and time within the field of physics, as part of general relativity. Space and time were no longer there *a priori*. On the contrary, the structure of space-time is linked to the distribution of matter, and Einstein gives us the tool for determining and studying that structure.

The incorporation of time and space within the theory was only a first step, since fundamental parameters still remain outside it. However, within string theory, fundamental parameters suffer the same fate time and space had suffered within the framework of general relativity. They lose their status as 'givens' of Nature, and become parameters capable of evolving. 'Constantology' could well become a discipline of physics, with Dirac as one of its founders.

## A PROMISING FUTURE

The commission has finally shed its concerns about the edifice of physics. It has even become rather optimistic about the century ahead. The situation now is very different from that of the late nineteenth century. Then, scientists thought they were on the verge of knowing and understanding everything: all that remained to do was to measure constants in Nature more accurately. Of course, some problems were proving intractable, but it was thought that they would soon be resolved. As we now know, these expectations were unrealised, and those few 'problems' carried the seeds of relativity and quantum mechanics, which revolutionised our view of Nature during the twentieth century.

Today, things are very different. Many questions remain open, especially those concerning the unification of fundamental forces, and the problems of dark matter and the cosmological constant. Moreover, it seems unlikely that these questions will be answered soon or succumb to a 'quick fix'. The solutions to these problems could reside in a complete revision of our current ideas. We might go as far as to replace particles by linear objects, increase the dimensions of space and envisage multiple universes...

The revolution in physics, if it comes, will be profound.

# Bibliography


Barrow, J.D., *New Theories of Everything,* Second Edition, Oxford University Press, 2007.

Barrow, J.D. and Tipler, F.J., *The Anthropic Cosmological Principle,* Oxford Paperbacks, 1988.

Calle, C., *Superstrings and Other Things; A Guide to Physics,* Institute of Physics Publishing, 2001.

Chown, M., *Quantum Theory Cannot Hurt You,* Faber & Faber, 2007.

Close, F., *The Void,* Oxford University Press, 2007.

Cohen-Tannoudji, G., *Universal Constants in Physics,* McGraw-Hill Horizons of Science Series, 1992.

Damour, T., *Si Einstein m'était conté,* Le Cherche-Midi, 2005.

Demaret, J. and Lambert, D., *Le Principe Anthropique – l'homme est-il au centre de l'Univers?,* Armand Colin, Paris, 1994.

Deutsch, D., *The Fabric of Reality: Towards a Theory of Everything,* New Edition, Penguin Books Ltd., 1998.

Diu, B. and Leclercq, B., *La physique mot à mot,* Odile Jacob, 2005.

Duff, M.J., Okun, L.B. and Veneziano, G., Trialogue on the number of fundamental constants, *Journal of High Energy Physics* 03023, 1-30 (2002).

Einsenstaedt, J., *Einstein et la relativité générale,* Editions du CNRS, 2002.

Feynman, R.P., *Six Not-so-easy Pieces: Einstein's Relativity, Symmetry and Space-time,* Penguin Press Sciences, Penguin Books Ltd., 1999.

Gamow, G., Stannard, R. and Edwards, M., *The New World of Mr Tompkins: George Gamow's Classic Mr Tompkins in Paperback,* Revised Edition, Cambridge University Press, 1999.

Greene, B., *The Elegant Universe: Superstrings, Hidden Dimensions, and the Quest for the Ultimate Theory,* New Edition, Vintage, 2005.

Greene, B., *The Fabric of the Cosmos: Space, Time and the Texture of Reality,* Penguin Press Science, Penguin Books Ltd., 2005.

Hawking, S., *On the Shoulders of Giants: The Great Works of Physics and Astronomy,* Penguin Books Ltd., 2003.

Lehoucq, R. and Uzan, J.-P., *Les constantes fondamentales,* Belin, 2005.

Lévy-Leblond, J.-M., 'The importance of being (a) constant', in: Enrico Fermi School LXXII, *Problems in the Foundations of Physics,* N. Toraldo di Francia and Bas van Fraassen (Editors), Amsterdam, North Holland, 1979.

Maxwell, J.C., Presidential address to the British Association for the Advancement of Science, 1870.

McMahon, D., *String Theory Demystified,* McGraw-Hill Professional, 2008.

Penrose, R., *The Road to Reality: A Complete Guide to the Laws of the Universe,* Vintage, 2006.

Petitjean, P., *Les raies d'absorption dans le spectre des quasars,* Annales de physiques, EDP Sciences, 1999.

Poincaré, H. and Bolduc, J.W., *Mathematics and Science: Last Essays,* Kessinger Publishing, 2007.

Rees, M., *Just Six Numbers: The Deep Forces that Shape the Universe,* Phoenix Science Masters, 2000.

Rosenthal-Schneider, I., *Reality and Scientific Truth: Discussions with Einstein, Von Laue and Planck,* Wayne State University Press, Detroit 1981.

Silk, J., *A Short History of the Universe,* Scientific American Library, New Edition, 1997.

Susskind, L., and Lindesay, J., *An Introduction to Black Holes, Information and the String Theory Revolution: The Holographic Universe,* World Scientific Publishing Co. Pte. Ltd., 2005.

Taten, R., *Histoire generale des sciences,* PUF, vol. 1957-1964.

Taylor B.N. (Editor), *The International System of Units (SI),* National Institute of Standards and Technology (NIST) Special Publication 330, 2001.

Weinberg, S., Overview of theoretical prospects for understanding the values of fundamental constants, in: *The Constants of Physics,* W.H. McCrea and M.J. Rees (Editors), Phil. Trans. R. Soc. London A 310, 249-252 (1983).

Weinberg, S., *The Discovery of Subatomic Particles,* Revised Edition, Cambridge University Press, 2003.

Woodhouse, N.M.J., *Special Relativity,* Springer Undergraduate Mathematics, Springer-Verlag London Ltd., 2007.

Woodhouse, N.M.J., *General Relativity,* Springer Undergraduate Mathematics, Springer-Verlag London Ltd., 2008.

# Index

Printing: Mercedes-Druck, Berlin
Binding: Stein + Lehmann, Berlin